Reforming New Zealand Secondary Education

SECONDARY EDUCATION IN A CHANGING WORLD

Series editors: Barry M. Franklin and Gary McCulloch

Published by Palgrave Macmillan:

The Comprehensive Public High School: Historical Perspectives
By Geoffrey Sherington and Craig Campbell
(2006)

Cyril Norwood and the Ideal of Secondary Education
By Gary McCulloch
(2007)

The Death of the Comprehensive High School?:
Historical, Contemporary, and Comparative Perspectives
Edited by Barry M. Franklin and Gary McCulloch
(2007)

The Emergence of Holocaust Education in American Schools
By Thomas D. Fallace
(2008)

The Standardization of American Schooling:
Linking Secondary and Higher Education, 1870–1910
By Marc A. VanOverbeke
(2008)

Education and Social Integration:
Comprehensive Schooling in Europe
By Susanne Wiborg
(2009)

Reforming New Zealand Secondary Education:
The Picot Report and the Road to Radical Reform
By Roger Openshaw
(2009)

Reforming New Zealand Secondary Education

The Picot Report and the Road to Radical Reform

Roger Openshaw

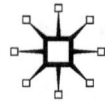

REFORMING NEW ZEALAND SECONDARY EDUCATION
Copyright © Roger Openshaw, 2009.

All rights reserved.

First published in 2009 by
PALGRAVE MACMILLAN®
in the United States—a division of St. Martin's Press LLC,
175 Fifth Avenue, New York, NY 10010.

Where this book is distributed in the UK, Europe and the rest of the world, this is by Palgrave Macmillan, a division of Macmillan Publishers Limited, registered in England, company number 785998, of Houndmills, Basingstoke, Hampshire RG21 6XS.

Palgrave Macmillan is the global academic imprint of the above companies and has companies and representatives throughout the world.

Palgrave® and Macmillan® are registered trademarks in the United States, the United Kingdom, Europe and other countries.

ISBN: 978-0-230-60626-5

Library of Congress Cataloging-in-Publication Data

Openshaw, Roger.
　　Reforming New Zealand secondary education : the Picot Report and the road to radical reform / Roger Openshaw.
　　　　p. cm.—(Secondary education in a changing world)
　　Includes bibliographical references and index.
　　ISBN 0-230-60626-1
　　　1. Education, Secondary—New Zealand. 2. Educational change—New Zealand. 3. Education and state–New Zealand. I. Title.
LA2126.O64 2009
379.93—dc22　　　　　　　　　　　　　　　　　　　　　　　　2009002651

A catalogue record of the book is available from the British Library.

Design by Newgen Imaging Systems (P) Ltd., Chennai, India.

First edition: September 2009

Transferred to Digital Printing in 2012

Contents

Series Editors' Foreword — vii
Acknowledgments — xi
Abbreviations — xiii

Part 1 Tensions and Contradictions

1. To Suit a Political Purpose? Reinterpreting the Educational Reforms — 3
2. Almost Alone in the World, 1942–1968 — 19

Part 2 Crises and Solutions

3. Game War — 43
4. Only Major and System-wide Reforms Will Suffice — 59
5. The Best Kind of Accountability — 75

Part 3 Elusive Consensus

6. A Blank Page Approach — 99
7. A Long Way to Go before We Win the Battle — 123
8. Only Half a Policy — 143

Conclusion: A Real Say or a National Morality Play? The Road to Radical Reform in Retrospect — 165

Notes — 185
Bibliography — 221
Index — 249

Series Editors' Foreword

Among the educational issues affecting policymakers, public officials, and citizens in modern, democratic, and industrial societies, none has been more contentious than the role of secondary schooling. In establishing the Secondary Education in a Changing World series with Palgrave Macmillan, our intent is to provide a venue for scholars in different national settings to explore critical and controversial issues surrounding secondary education. We envision our series as a place for the airing and hopefully the resolution of these controversial issues.

More than a century has elapsed since Emile Durkheim argued the importance of studying secondary education as a unity, rather than in relation to the wide range of subjects and the division of pedagogical labor of which it was composed. Only thus, he insisted, would it be possible to have the ends and aims of secondary education constantly in view. The failure to do so accounted for a great deal of the difficulty with which secondary education was faced. First, it meant that secondary education was "intellectually disorientated," between "a past which is dying and a future which is still undecided," and as a result "lacks the vigor and vitality which it once possessed" (Durkheim 1938/1977, p. 8). Second, the institutions of secondary education were not understood adequately in relation to their past, which was "the soil which nourished them and gave them their present meaning, and apart from which they cannot be examined without a great deal of impoverishment and distortion" (p. 10). And third, it was difficult for secondary school teachers, who were responsible for putting policy reforms into practice, to understand the nature of the problems and issues that prompted them.

In the early years of the twenty-first century, Durkheim's strictures still have resonance. The intellectual disorientation of secondary education is more evident than ever as it is caught up in successive waves of policy changes. The connections between the present and the past have become increasingly hard to trace and untangle. Moreover, the distance between policymakers on the one hand and the practitioners on the other has rarely seemed as immense as it is today. The key mission of the current series of books is, in the spirit of Durkheim, to address these underlying dilemmas of secondary education and to play a part in resolving them.

Roger Openshaw's study, *Reforming New Zealand Secondary Education: The Picot Report and the Road to Radical Reform*, traces in detail the radical reforms proposed in New Zealand by the Picot Report, *Administering for Excellence*, in 1987, and implemented the following year. It documents the gathering impetus for change in the education system over the previous thirty years arising from a range of political and social developments. It investigates the influences on the Picot Report from the Treasury and neo-liberal pressure groups, and also from educators and others who were concerned to promote greater social equity and improve the achievement of underperforming schools and pupils. It shows the efforts made by the then Labour government under David Lange to sell the reforms to educators and the wider public, including the unprecedented use made of a public relations firm to support its communications strategy. It follows the implementation of the reforms, and the compromises that were involved in this over the years that followed. Finally, it assesses the impact of the reforms on education in New Zealand over the past twenty years and into the present. In all of this, it seeks to evaluate the rival claims of the protagonists in a debate over education that was at the heart of New Zealand's political and social struggles in the turbulent years of Lange's Labour government.

New Zealand's educational reforms were closely related to the worldwide movement for policy change, including fundamental reforms put in place in the United States, the United Kingdom, and elsewhere. This book makes an important contribution to an understanding of the global movement for educational reform in these years. Yet New Zealand's reforms were also distinctive in many ways, and Openshaw's account draws out these specificities very clearly. For example, the educational reforms in New Zealand, unlike those in many other countries, placed a key emphasis on community issues and ideals. Advocates for the indigenous Māori community also played a significant role in determining the character of these reforms. The changes that arose were to affect all areas of education in New Zealand, not only secondary education. Yet Openshaw also shows that in many respects the source of the complaints that led to the Picot report lay with the secondary schools, and that it was to affect them and their teachers and pupils in fundamental ways that continue to this day.

Openshaw's book analyzes a very large amount of documentary evidence of the reforms that has never been researched or published in any previous account. These allow him to demonstrate the Picot committee's aim to begin with what it called a "blank page" approach, on which it tried to erect a new administrative structure that would be more efficient and effective than the old system which had been in existence for over a hundred years. He is also able to expose to critical scrutiny the myths that grew up around Picot from across the political spectrum. His analysis will be essential reading not only for historians and educators, but also for policy analysts and students of politics—and not only in New Zealand, but wherever educational reforms have been tried and often failed over the past generation.

Reforming New Zealand Secondary Education is the seventh volume to be published in our series. It continues and develops further our key project of promoting an enhanced understanding of the international and indeed global context within which secondary education has developed. It also pursues the theme of policy reform that has been prominent throughout, and with it the issue of how far such reform has provided a coherent basis for how secondary education will be understood and practiced in the twenty-first century. As we see the trajectory of the series advancing during the next few years, we hope to support further work that brings these broad and fundamental concerns to studies in secondary education.

Barry Franklin and Gary McCulloch
Series Co-editors

Reference

Durkheim, E. (1938/1977). *The Evolution of Educational Thought: Lectures on the Formation and Development of Secondary Education in France.* London: Routledge and Kegan Paul.

Acknowledgments

Most research projects owe their completion to the hard work and dedication of many. This particular project has been no exception. The author is grateful to the following institutions and individuals for their assistance in bringing this book to fruition. Archivists at both the Wellington and Auckland branches of Archives New Zealand freely and consistently furnished both knowledge and expertise in locating and accessing relevant material. Prior to his death, the late David Lange kindly permitted me access to education material contained within the Lange papers held at Archives New Zealand (Wellington). At both the National Library and Alexander-Turnbull Library, library staffs were most helpful in responding to my numerous requests for assistance. I would like also to thank the general staff of National Library, Archives and the James Cook Grand Chancellor, Wellington, for being so understanding in accommodating my frequently bulky overnight luggage, complete with my portable office including laptop, separate keyboard, and mouse. At Massey University, librarians at Turitea and at Hokowhitu promptly and expertly dealt with my particular research needs.

On an individual level, I would like to thank a number of close colleagues both in New Zealand and the United Kingdom who gave up their time both to listen and to offer comment on my research, and who pointed me in the right directions. This book began when I was working with Dr. Janet Soler at the Open University on another project, and I am grateful to her for urging me to proceed with it. The book's appearance in this timely series on secondary education worldwide is due to the encouragement given to me by Professor Gary McCulloch, Brian Simon Professor of the History of Education, Institute of Education, University of London.

A number of friends and colleagues here in New Zealand have been instrumental in furthering my research. I will name just a few of them here. Professor Howard Lee, my Head of School, has been particularly understanding and helpful. He never failed to offer me excellence guidance and to share his wide historical knowledge with me, despite an extremely high workload. I have also benefited from many conversations with Associate

Professor John Clark, who provided numerous research insights for which I am extremely grateful. Both Associate Professor Margaret Walshaw at Massey and Associate Professor Elizabeth Rata at the University of Auckland were particularly patient listeners and supportive colleagues whose suggestions benefited this book. I should also to record my gratitude to the secretarial staff in the School of Educational Studies at Massey University College of Education, particularly Ms Jenny Rive who made my many travel arrangements, and Ms Tracey Beattie-Pinfold who took so much time and trouble to format and edit the manuscript prior to its being handed over to the publisher.

Last but by no means least, I would like to thank my family; my three sons, Jonathan, William, and Richard, and my partner, Margaret, for their understanding and forbearance with an at times twitchy author.

Note: The text conforms to U.S. spelling and conventions throughout, except in the case of direct quotations, which have been left in the original spelling, UK or U.S. as the case may be.

Abbreviations

AAFH	State Services Commission Files
AALR	Treasury Files
AAWW	David Lange Papers
AAZY	Departmental Residual Management Files Unit
ABEP	Department of Education and Ministry of Education Head Office Files
ABRP	Wanganui Office of Te Puni Kokiri
ACEP	Advisory Council on Educational Planning
APEC	Asia-Pacific Economic Cooperation
asTTle	Assessment Tools for Teaching and Learning
BAAA	Northern Regional Office
BCDQ	Regional Department of Education Residual Management Unit Files
BoT	Board of Trustees
CARE	Citizens Association for Racial Equality
CDA	Critical Discourse Analysis
CDU	Curriculum Development Unit
CEIS	Community Education Initiative Scheme
CEO	Chief Executive Officer
COPE	Committee of Officials on Public Expenditure
CPA	Concerned Parents Association
EDC	Educational Development Conference
ERO	Education Review Office
FFS	A Computer Forecasting System
GNP	Gross National Product
IEA	Institute of Economic Affairs
MACOS	United States–developed social studies programme
MoE	Ministry of Education
MP	Member of Parliament
NACME	National Advisory Committee on Māori Education
NCEA	National Certificate of Educational Achievement
NCEA	National Certificate of Achievement

NDC	National Development Council
NRO	National Regional Office
NZCER	New Zealand Council for Educational Research
NZCF	New Zealand Curriculum Framework
NZCF	New Zealand Curriculum Framework
NZJES	New Zealand Journal of Educational Studies
NZPD	New Zealand Parliamentary Debates
NZQA	New Zealand Qualifications Authority
NZQF	New Zealand Qualifications Framework
NZSTA	New Zealand School Trustees Association
OECD	Organisation for Economic Co-Operation and Development
PPTA	Post-Primary Teachers' Association
PR	Positions of Responsibility
RAA	Review and Audit Agency
SES	Special Education Services
SSC	State Services Commission
STA	School Trustee's Association
STEPS	School-Leavers Training and Employment Preparation Scheme
TRB	Teacher Registration Board
YCDB	Auckland Education Residual Management Unit
YPTP	Young Persons Training Programme

Part 1

Tensions and Contradictions

Chapter 1

To Suit a Political Purpose? Reinterpreting the Educational Reforms

In 1877 New Zealand created a three-tiered public education system consisting of a centralized Department of Education, regional education boards, and local school committees. Apart from some limited provision for scholarships to District High Schools, secondary education was to be reserved for a fee-paying colonial elite. Some 110 years later, in July 1987, a taskforce headed by Brian Picot was charged by Labour prime minister and soon to be minister of education, David Lange, with comprehensively reviewing administrative efficiency across a greatly expanded education system that now incorporated early childhood, primary, secondary, and post-compulsory education. Its brief included a critical scrutiny of the Department's functions along with a reassessment of school, college, and polytechnic governing bodies, with a view to delegating responsibilities and increasing community control.[1] Some nine months later, in April 1988, the taskforce's report, *Administering for Excellence*, was released. It concluded that the structure of the existing education system

> is a creaky, cumbersome affair. It is not the result of an overall plan or design, but has taken on its present shape by increments and accretion. Such a haphazard collection of administrative arrangements is not suited to the rapidly changing late twentieth century. In looking at the system, we have observed a number of serious weaknesses. These can be grouped

broadly under the following themes:
- overcentralisation of decision making
- complexity
- lack of information and choice
- lack of effective management practices
- feelings of powerlessness[2]

Adopting a "blank page" approach to educational reform, the taskforce proposed an entirely new structure for education.[3] All schools were henceforth to be autonomous, self-managing learning institutions, controlled by locally elected boards of trustees responsible for learning outcomes, budgeting, and the employment of teachers. Each learning institution was to produce a charter outlining the school's mission in relation to its clientele and community, incorporating centrally prescribed requirements of safety, legality, equity, and national standards.[4]

The taskforce recommended the abolition of the Department of Education, the regional education boards, and the local school committees/boards of governors in favor of a simplified two-tiered administrative system. There was to be a new Ministry of Education consisting of three major divisions—policy, property, and operations. An Education Review Office (ERO) was to be set up with responsibility for reporting on all schools and centers, including the provision of critical advice on school performance and practices. The taskforce also recommended the creation of a Parent Advocacy Council and Community Education Forums, both of which were subsequently abolished as retrenchment measures under a newly elected National government in 1991. In August 1988, some four months after *Administering for Excellence* had appeared *Tomorrow's Schools*, the government's response to the Picot Taskforce's recommendations was released. This document largely accepted the recommendations of the Picot Report. Legislation to give effect to the promised new era in education came into force on October 1, 1989.

Over the next two decades the changes recommended in the Picot Report were to encompass nearly every aspect of New Zealand education. Not surprisingly, the motives for the changes together with their perceived impact on schools and communities have received intense scrutiny from within and outside the sector. One reason for this sustained interest is that, although the Picot Report and *Tomorrows' Schools* focused largely on administrative issues, their ultimate significance lies in the fact that systems of educational administration are conditioned by the prevailing political and educational philosophies of the societies in which they operate, as Sir Frank Holmes once perceptively observed.[5] Although neither document focused exclusively on secondary education, the underlying motives for the reforms, the policy discourse invoked to justify them, as well as much of the subsequent debate about educational outcomes centered largely on public concern over state secondary schools; the institutions that were for most New Zealand students in this period, the exit point for paid employment or post-compulsory education.

The Reforms and the Research Polarization

Broadly speaking, there are two diametrically opposed views on the motives and overall impact of the reforms on schools, teachers, parents, and students. Their significance lies partly in the fact that each has particular consequences for the way we view the history, present operation and future direction of secondary schooling. For those with a particular interest in secondary education history, there are also ongoing research implications regarding the questions we pose, the theoretical tools we deploy, the evidence we present, and ultimately how we position the New Zealand reforms within a wider global context of educational change.

The first view of the reforms, held by a number of commentators located largely but not exclusively outside of the education sector, largely accepts the Picot Taskforce's assertion that the old education structures were both outdated and inflexible, particularly with regard to secondary education. The system is seen as having encouraged provider capture to the extent that by the mid-1980s it had long ceased to operate in the best interests of either the country as a whole, or of the public at large. Consequently, the Picot Taskforce's recommendations were a legitimate expression of a growing public disenchantment with education. Simon Smelt, chief analyst with Treasury, formerly manager, Treasury Education Section and Treasury representative to the taskforce, was an early promoter of such a view. In a book indicatively titled, *Today's Schools. Governance and Quality*, Smelt hailed the initial implementation of the reforms as "a unique and bold attempt to counter the perceived problems of the previous structure by abolishing layers of administration and empowering parents."[6]

This praise for the perceived intentions of the reformers, however, was far from being an uncritical endorsement of what followed. Smelt conceded that the reforms had "taken time to bed down" and that some elements had "developed in unforeseen ways."[7]

> Looked at in terms of the four issues of cooperation, conflict, co-ordination and capabilities, the reforms have resolved some problems but others have emerged or been revealed. Underlying problems, such as political pressures and the risk the state holds in respect of the school system, have themselves

imposed limitations on the extent of delegation. Because of the continuing Crown stake in, and exposure to, schooling, delegation is far from complete and the agents of the state have attempted to minimise risk and ensure appropriate schooling through input controls, contract specification and enforcement, and a focus on ensuring correct process.[8]

In Smelt's view the most effective way of countering the efforts of these agents of the state, represented by the former education bureaucracy and the teachers unions, lay in strengthening the original pillars of the Picot reforms. This could be achieved by: switching the emphasis on input controls to the specification of minimum outputs to be achieved; by introducing better procedures to deal with failing schools; by improving support for boards of trustees including the encouragement of inter-board cooperation, and through the encouragement of diversity that might in turn lead to greater choice for educational consumers.

Smelt's book furnished no fanfare to a Treasury triumph. On the contrary, he warned his readers that the best intentions of the reformers had been in some respects, subverted. Smelt drew attention to the growing gap between the original intentions of the Picot Taskforce and the subsequent modifications to its recommendations—a discrepancy that a number of right-wing and business critics of the reforms were to subsequently identify. Commenting on the criticisms made of New Zealand education by the Sexton Report, the Porter project, and the World Competitiveness Report, the executive director of the New Zealand Business Roundtable, Roger Kerr, conceded that the Picot Report had made a useful advance in arguing for decentralization of educational decision-making within a framework of national objectives. However, Kerr also identified what he considered to be a number of serious weaknesses. Picot had focused exclusively on the management of schools by parent representatives, but Kerr felt that parents should not necessarily have to run schools to obtain good education for their children. Further

> the Picot Report did not judge it politically expedient to explore some of the normal ways of satisfying consumer demands—through choice and competition in the market. Moreover, as the Sexton Report documents, the Picot Reforms were watered down in their implementation in the face of opposition from the education establishment.[9]

Once again, we see a common theme in right-wing critiques that dwells not on the success, but on the failure of the reforms to fulfill their early promise. Kerr was particularly critical of what he saw as a continuing propaganda campaign emanating from the teachers unions and some academics within education aimed at convincing the public that, since Picot, the system had been effectively captured by so-called New Right views that emphasized technocratic, meritocratic, and competitive views of education. Kerr, however, believed that the post-Picot education system was still a long way from enshrining such values in its schools.[10]

Writing some sixteen years later National member of parliament, Allan Peachey, a prominent ex-principal of Rangitoto College, one of the country's largest secondary schools, was to share much of Kerr's enthusiasm for the original intent of the reforms while lamenting that so little had actually been achieved. Often credited with inspiring dramatic improvements that brought his secondary school's academic and sporting achievements into line with its community's high decile rating, Peachey believed that a major gain of the education reforms had been a heightened awareness of the significance of "excellence," stimulating increased community pride in the school and leading to improvements in teaching and learning.[11] Like Kerr, however, Peachey argued that the Picot Report had failed in its ultimate purpose of encouraging all schools to become excellent, largely because the original intentions of the reforms had been subverted by subsequent developments. Worse, a lack of political will on the part of successive governments had failed to curb the power of teacher unionism. The result was that powerful education bureaucracies had reemerged to wrest back the control of education from local communities. Accordingly, "honest educational thinkers (would), in time, look back on the 1990s and conclude that a marvellous opportunity was wasted as a consequence of political expediency and trade-union self-interest."[12]

Smelt, Kerr, and Peachey were by no means alone in adopting a critical view of the Taskforce's impact. A second view of the reforms; one held by many critical theorists specializing in education policy, shared their concern that the original high hopes of the New Zealand public and the parents had been dashed by those determined to mould the reforms in their own images. This second view, however, was deeply suspicious of the motives underpinning the reforms, perceiving the outcomes to be far worse than any that had preceded them. For some critics, this entailed a somewhat ironic reversal of their previous position that the former New Zealand education system had largely served hegemonic interests, in favor of at least a partial return to the previous liberal-progressive depiction of a system that had served the nation reasonably well, and enjoyed considerable public confidence. The Quality Public Education Coalition, launched in October 1997 with the expressed aim of combating what it saw as an unacceptable upsurge in the influence of market forces and competition upon schools, actively promoted the view that "people need to know the history of public education, which is that it was put in place to deal with the problems created by a dog-eat-dog society."[13]

It followed from this revised position that, in the absence of widespread public demand for change, the Picot Taskforce should be regarded not as a natural or inevitable development of parental wishes, nor as an expression of popular democracy in action, but rather as an unwitting puppet of a largely imported neoliberal ideology. In a critique that drew heavily upon contemporary critical theory, a growing number of education policy researchers argued that Treasury captured the Picot Committee along with much of the state's decision-making apparatus through a cynical exploitation of a wider crisis in capitalism, leading inevitably to the progressive dismantling of the old social-democratic social service sector, including education.

Instrumental in promoting this view was a two-part *New Zealand Journal of Educational Studies* (*NZJES*) article that appeared in 1988. This article highlighted the profound and detrimental impact on education of the New Zealand Treasury's *Government Management: Brief to the Incoming Government* (1987), contending that this document initiated a "Third Wave" of key reforms that effectively reversed the century-old expansion of universal education.[14] Some three years after the appearance of this influential critique, a collection of papers by key education policy researchers argued that the Picot Report and the subsequent implementation of *Tomorrow's Schools* had transformed the public education system for the worst.[15] The culprits in this disastrous exercise were Treasury and its "New Right" ideologues.[16] Far from being an ad hoc collection of well-meaning reformers, the New Right was in fact a closely knit political entity, promoting "a family of mutually consistent concepts," with Treasury being "the political thought-police of the New Right revolution."[17]

Ivan Snook epitomized those researchers who drew upon critical theory to promote the concept of a sharp break between pre- and post-reform periods, identifying a dichotomy between "introduced" New Right theory and "indigenous" New Zealand beliefs and values.[18] Gerald Grace endorsed and extended the idea that there was a yawning chasm rather than a mere gap between pre- and post-reform eras. Grace praised the educational settlement created under the first Labour government that spawned the early postwar secondary school comprehensive ideal, going on to contrast this sharply with post-1987 developments.[19] Liz Gordon took this dichotomy still further in her claim that post-1987 developments effectively reversed policies built on a liberal-democratic consensus that successive governments in New Zealand had supported—a consensus "pivoted on the understanding that education should not be 'political' or partisan, but should be seen as a public good to which all should work with equal commitment."[20]

Behind this dramatic reversal of traditional policy, it was claimed, lay fundamental changes in the economy and the role of the state. Codd,

Harker, and Nash[21] argued (following Offe), that the policy of devolution initiated by Picot constituted a strategic response to the political and economic crises faced by Western capitalist nations during the mid-1980s.[22] Roger Dale epitomized this position with his contention that the Keynesian welfare state became exhausted due to a mixture of intrinsic contradictions and changing socioeconomic conditions.[23] In seeking to explain just why it was that the Picot-inspired educational reforms found widespread public acquiescence, critical theorists largely drew upon U.S. and British research that purported to explain how it was that conservative ideologies were able to attract diverse groups of supporters who would not ordinarily have allied themselves with market policies in education. Thus according to Michael Apple,

> the first thing to ask about an ideology is not what is false about it, but what is true about it. What are its connections to lived experience? Ideologies, properly conceived, do not dupe people. To be effective they must connect to real problems, real experiences... the movement away from social democratic principles and an acceptance of more Right Wing positions in social and educational policy occur precisely because conservative groups have been able to work on popular sentiments, to reorganize genuine feelings, and in the process to win adherents.[24]

The passage of time has, if anything, led to a hardening of attitudes toward the whole era of public sector reforms initiated by Lange's Labour Government. Two *Delta* volumes appearing in the late 1990s edited by Anne-Marie O'Neill aptly conveyed the strong sense of betrayal felt by those with left-wing sympathies, many of whom had originally welcomed the Labour's 1984 election victory.[25] A growing sense of powerlessness was also evident. Thus, in his introduction to a special issue of the *New Zealand Journal of Educational Studies*, Martin Thrupp identified the major concern behind most of the articles in the volume as being the pervasiveness of neoliberalism or economic rationalism.[26]

In this same volume Patrick Fitzsimons, Michael Peters, and Peter Roberts, drawing upon the work of U.S. and European critical theorists, epitomized the continuing conviction amongst critical theorists that that British and American commitment to monetarism and supply-side economics provided the global context for copycat neoliberal structural reforms in New Zealand. The New Zealand public policy regime had been captured by a "scientized" economic discourse, represented in education by a Picot Report that was from the outset driven by (in)famous Treasury statements about the dangers of provider-capture. In turn this had led to "teacher-proofing" the curriculum, representing a decisive break with past educational traditions based on collaboration and consensus.[27] A related consequence of the reforms according to John Codd was the implementation of what he terms "policies of distrust," embodied in the separation of policy formation and policy advice from policy implementation, and in the related separation of funder from provider.[28] Jonathan Boston went even further, arguing that "within the OECD, probably the most radical and comprehensive programme of state-sector reform has been pursued in New Zealand."[29] These changes, Boston claimed, were stimulated by perceptions of bureaucratic shortcomings, high inflation, rising unemployment, large fiscal deficits, mounting public debt, and slow economic growth:

> Broadly speaking, these changes have conformed to the principles and practices of managerialism. For instance, there has been a strong emphasis on the devolution of management responsibilities.... The privatisation of commercial state assets, the commercialisation of many departmental functions, the institutional separation of public funding and provision, and the separation of the functions of policy advice, regulation and policy implementation. These changes have not only brought about a radical reshaping of the bureaucratic landscape, but also contributed to a dramatic "downsizing" of the core public sector and the development of a new managerial ethos.[30]

More recently, critical theorists have attempted to outline some possible strategies of resistance in the face of what they see as an all-pervasive global neoliberalism. Mark Olssen has argued that the main task of critical theorists lay in "exposing the contradictions of New Right policies in education,"[31] the ultimate aim being influencing policy through collective political action. Some indication of where such strategies might lead have been subsequently discussed by Olssen, Codd, and O'Neill.[32] In maintaining the position that there was indeed a sharp dichotomy between the pre- and post-Picot eras, Olssen et al focus particularly on the way the New Zealand state deliberately chose to emulate the Anglo-Saxon example epitomized by Reagan's United States and Thatcher's Britain in reacting to the imperatives of globalization through the strategy of deregulation.

> Central to our argument, then, is the claim that it is imposed policies of neoliberal governmentality, rather than globalization as such, that is the key forces affecting (and undermining) nation-states today. Thus, while a great deal of recent educational policy can be explained in terms of the sociological concept of globalization, we argue in this book that it must be theoretically represented in relation to the political philosophy of neoliberalism.[33]

Far from being an end in itself, this critique was seen to provide the basis for future political action. In the view of Olssen et al., neoliberalism incorporates an impoverished conception of the individual, human nature, and state power, hence they argue for a new model of community located not in the celebration of individual enterprise and initiative, but rather within a communitarian context within which the state would, even more than in welfare states of the past, become an education state.[34]

The Research Legacy: Challenge and Response

The continuing polarization of views on the reforms has continued to cast a lengthy shadow on historical research into education policy in New Zealand. In some quarters this has resulted in a reluctance to acknowledge the recent educational past at all, through fear of causing offence to one group or another. Typical of this desire to move on from past debates is Jane Gilbert's evocatively titled *Catching the Knowledge Wave* which advocates further reform of secondary schooling in order to make it more relevant to the demands of the emerging global society. As a postmodernist Gilbert recognizes the necessity of understanding and deconstructing discourses as a key to social change, but she largely accepts the view that New Zealand now has a more accountable education system as a result of the reforms ushered in by Picot and by *Tomorrow's Schools*. Hence, successful secondary education is viewed largely as marrying the successful teaching of key competencies with improved equal opportunity policies in order to increase the nation's ability to compete on the world economic stage as a high-skills, high-wage economy.[35] This view seems quite close to the position currently espoused by the New Zealand Ministry of Education. The ministry, however, clearly operates within a post-Picot administrative structure. Consequently, the dangers in simply buying into the new rhetoric can be readily discerned in the way the ministry continues to market its underlying educational philosophy as one of "rigorous eclecticism, leaving any underlying contradictions to go largely unchallenged."[36]

Perhaps, however, a degree of historical amnesia is understandable. In 1998, a book commissioned by the New Zealand Ministry of Education, written by independent historians Graham and Susan Butterworth provoked a sharp reaction from those within the education sector. The Butterworths tended to see the Picot Report as the outcome of widespread pressure for educational decision-making to be wrested from an educational bureaucracy increasingly out of touch with the realities of contemporary New Zealand society. In attempting to answer the potentially thorny question of whether the reforms actually constituted a "famous victory," however, they gave only a qualified "yes," conceding only that, for better or worse, the Picot Report and its subsequent implementation into legislation had thrust the country into a new era.[37] The Butterworths were aware that, particularly within educational circles, a considerably less charitable view of the reforms was extant. Hence, in arguing that the recent educational reforms were the inevitable outcome of deep-seated historical processes, they were in effect also issuing a direct challenge to received views among educators that the reforms had been largely driven by Treasury.[38]

Not surprisingly, the new book provoked an immediate and hostile response from some educational researchers. New Zealand educational historian David McKenzie, characterized the research as one of the new brand of so-called instant histories written with the principal purpose of promoting the newly created state ministries. In his view the unwillingness of the authors to utilize a readily available body of critical policy literature that challenged the received view that the reforms had been largely beneficial, simply reflected the official conviction that "most of the academics in university education departments were left-wing neo-Marxists uselessly in love with their own dogmatics."[39] In response the Butterworths charged that "as few of the academic commentators of the past decade ha[d] dirtied their hands with primary research on the development of government policy, they had little to contribute."[40] This emphatic dismissal of historical research emanating from university education faculties and colleges provoked an equally pointed counter-response. Effectively reversing the charge the Butterworths had laid against educational historians, Greg and Howard Lee pointed out that *Reforming Education* had blithely ignored the availability of a substantial body of literature emanating from policy analysts within university education departments. Furthermore, in claiming to be above and beyond debate, the Butterworths had equally ignored the reality that any study of education policy was, by definition, "political."[41]

To some degree this sharp exchange recalled similar episodes that had occurred in the United Kingdom some decades earlier, when a number of historians domiciled within established university history departments dismissed the history produced in university education departments as "impoverished and unsophisticated."[42] As early as 1976, the conservative historian Sir Geoffrey Elton denounced teachers colleges as strongholds of a heavily deterministic leftist history, produced by educationalists whose main interest lay in ransacking the past in the interests of supporting the radical reform of contemporary schooling.[43] A similar critique from a very different ideological perspective emanated from Brian Simon, a prominent Marxist educational historian, who warned that overly simplistic determinism had seriously undermined the Left's intellectual credibility.[44] Moreover, in 1983 Harold Silver, another prominent British educational historian, further contended that much recent educational history written in the United Kingdom had been overly polemical in articulating a strongly left-wing ideology, resulting in a further marginalization of curriculum history in the eyes of policymakers, the public, teachers, and other academics.[45]

It should be emphasized, however, that in New Zealand at least, the long-standing controversy surrounding the educational reforms of the 1980s has not been confined to those writing history in education faculties. Paul Goldsmith's review of Malcolm McKinnon's recent history of the New Zealand Treasury critiqued the author's coverage of Treasury policy in the crucial decades of the 1970s and 1980s for failing to give an adequate insight into the increasingly fierce internal battles that raged *within* Treasury from the late 1970s, and neglecting the deep intergenerational divide over economic and social policy within the country at large. Instead, wrote Goldsmith, McKinnon "appears to have swallowed the line put about by Prime Minister David Lange, post 1987, that "the Treasury agenda" was all pervasive and powerful."

> The ascription to the Treasury of the key role was Lange's way of deflecting blame for his own role in the reforms. It was also a way for critics from the left to rationalize what seemed to be traitorous actions by their party. Instead they were victims of a conspiracy. However, the evidence seems to be to the contrary. The striking aspect is the extent to which all the major transformations in New Zealand's fiscal and monetary policies, and the decisions not to transform, were driven by the politicians.[46]

One response from historians to the ongoing controversy over the entire era of social and economic reform in New Zealand ushered in by the Lange Government was to emphasize the significance of ongoing historical processes in stimulating and shaping educational reform. Thus David Thomson looked well beyond education as such, to attribute the growing disillusionment that swept Labour to power in 1984 to intergenerational conflict, as a younger generation discovered that the promises of the welfare state, including education, had worn threadbare.[47] Colin James took an even broader perspective in arguing that "the cycle between education and the economy was neglected in favor of a security that neglect helped make illusionary through failing economic performance."[48] Thus, in claiming that from the 1970s educational results increasingly failed to live up to expectations, James came to emphasize that "what was going wrong was not in the output from educators, but in the reception by students and society generally."[49]

Within the education sector there has also been some willingness to examine more closely the complex interplay of historical forces that make for radical change. According to Snook, for instance

> the Treasury took seriously all the "leftish," "radical[,]" "sociological[,]" critiques of education, all of which show that in terms of fostering equality, the education system...has failed. It then turned the argument on its head, holding that a "radically" new solution was required, not (as liberals think) a tinkering with the existing system.[50]

This view, however, still placed Treasury as the sole active agent in a neoliberal conspiracy to destroy the supposedly social-democratic consensus that had previously underpinned New Zealand education. Historians have enjoyed only limited success in challenging this view.[51] John Barrington was one of the first to do so, claiming that the fundamental changes in the administration of primary and secondary schools that took place after the election of Labour in 1984 had been inadequately explained in terms that placed a heavy emphasis on the activities of the contemporary "New Right." In so doing, critics had overlooked to an extraordinary degree the "historical developments which influenced and helped create the preconditions for change."[52] Gary McCulloch was another historian who opted for a more critical approach to causation. In particular McCulloch questioned the validity of tightly defined criteria in assessing complex political categories such as neoliberalism, arguing that single-cause explanations centering exclusively on the impact of imported neoliberal ideals tended to ignore the cumulative impact of indigenous historical factors.[53] Consequently, the New Right was depicted as an essentially alien importation into a land that, prior to the 1980s, had largely experienced democratic consensus in education. My own 1995 study of post-primary education utilized the concept of educational settlements to demonstrate how common interests amongst otherwise ideologically opposed groups had underpinned major changes in the way education was conceived of and organized. Thus, by 1984 a combination of social, economic, educational, and political changes combined to produce a policy environment where radical changes were being advocated across a wide section of society. The outcome was an educational reform package that reflected an ideologically mixed heritage.[54]

A better understanding of the rapidly changing education policy environment during the entire post–World War Two era, with a specific concentration on the decade immediately prior to the setting up of the Picot Taskforce, seemed to be called for. Such an approach necessitated a renewed focus on primary source material, taking into account McCulloch's recent plea for the need to understand documents in relation to their milieu—in short, to relate text to context.[55] A first step was to look at the steady build-up of dissatisfaction with state education beginning in the 1940s with opposition to the comprehensive, multilateral state secondary school, and culminating in the decade immediately prior to the setting up of the Picot Taskforce, when a combination of Māori radical groups, feminists, radical academics, conservative pressure groups, cost-conscious politicians, and Treasury combined to demand radical educational change.[56]

This historical view, while still making but limited headway in the face of now entrenched views on the reforms at least suggested that the Picot Report might be situated at the end of a continuum, rather than at its beginning. However, in raising further questions about the complex interplay of ideologies at the policymaking level, it underlined the need for a more finely detailed, historically informed reexamination of late twentieth century New Zealand educational policymaking both as a variation of international developments and as the outcome of political and cultural forces that were distinctively national in character. Such a model might well have to recognize that Treasury was not the only influential actor on the educational policymaking stage. More ambitiously perhaps, the outcome might be a first step toward the replacement of an essentially unidimensional model of public policy with a more dynamic multidimensional model that situated educational reform within the wider context of public policymaking.

The New Acquisitions and the Potential for Radical Revision

A great deal has been written about the Picot Report and *Tomorrow's Schools* both at the time they were released, and subsequently. In addition to the academic and professional literature available, there have been countless newspaper and magazine articles and editorials of admittedly varying quality and usefulness.[57]

Moreover, the reforms were and occasionally continue to be, the subject of extended parliamentary debate over the past two decades. Once the reforms were in place, successive governments were forced to justify any planned changes to the legislation in the face of Opposition hostility, often leading to vitriolic exchanges in the House. Comparatively early on in this research, I opted to widen its scope beyond education to encompass the various reports, commissions and inquiries into other public sector agencies and activities released during the decade immediately prior to the setting up of the Picot Taskforce. Given that this book is designed to fit within a series dealing with the global evolution of secondary education, clearly something had to be left out. Largely for reasons of space, I decided to omit discussion of the near-contemporary Meade Report (on early childhood education), and the Hawke Report (on tertiary education). I also acknowledge that much remains to be written about post–*Tomorrow's Schools* developments, particularly in regards to curriculum and assessment. Once again for reasons of space, I reluctantly chose to deal with the New Zealand Curriculum Framework (NZCF) and the controversy that surrounded it, only in passing. As with the Picot Report, recently available primary source material will now made it possible to research in some detail the fascinating history of what became an extended curriculum debate, particularly over the first NZCF draft, which paralleled debate over the wider education reforms during a broadly similar period. Unfortunately though, this must be reserved for another time and place.

It will be obvious by now that the writing of this book has been shaped by my growing conviction as a university teacher and researcher, that many of our perceptions of the reforms have been shaped more by the continuing debate over them, than by hard evidence from the reform period itself or from the decades that preceded it. Indeed historical bias is so difficult to eliminate that some historians have given up the task altogether; hence Keith Jenkins has claimed that "all history is theoretical and all theories are positioned and positioning."[58] However, Richard Evans, in a statement that might well have been written for the extended debate over the New Zealand educational reforms has warned that

> ultimately, if political or moral aims become paramount in the writing of history, then scholarship suffers. Facts are mined to prove a case; evidence is twisted to suit a political purpose; inconvenient documents are ignored; sources deliberately misconstrued or misinterpreted. If historians are not engaged in the pursuit of truth, then scholarly criteria becomes irrelevant in assessing the merits of a particular historical argument.[59]

Thus, historians of the reform era are left with little recourse but to paradoxically pursue what Richard Dawkins has recently described as "the honest and systematic endeavour to find out the truth about the real world,"[60] while accepting the ultimate impossibility of that task. Accordingly, one way forward in researching the reforms seemed to be through a more comprehensive exploration of hitherto unaccessed primary source material. To be fair to those who have previously written about the Picot Report and *Tomorrow's Schools*, detailed information concerning the day-by-day deliberations of the Picot Taskforce and the material presented to it, both written and oral, has only become progressively available since the mid-1990s. Much the same can be said of the various state sector responses to its recommendations, and the lengthy but revealing internal struggle over the new educational structures.

In August 1995 Archives New Zealand acquired what was then the largest single accession in its history, emanating from the former Department of Education's Wellington Head Office (ABEP). Consisting of approximately 3,828 boxes of files, this accession provides a particularly comprehensive picture of Departmental reactions to curriculum, financial and administrative policy over the final decade of the Department's existence. Since then, further new archival acquisitions have become accessible that collectively have the potential to revolutionize research into the Picot era. Archives New Zealand has recently received some thirty boxes of Picot-related material under the general descriptor, Departmental Residual Management Files Unit (AAZY). Space precludes a full description but some indication of their importance can be illustrated by reference to a single box—enigmatically labeled "box 181." This box was discovered to contain the minutes to the first two crucial meetings of the Picot Taskforce in July and August 1987 respectively. It also contained several other particularly informative documents. One of these consisted of a large number of abstracts that furnished detailed summaries of the reading made available to the taskforce. Another turned out to be a handout to committee members on the subject of education vouchers, produced by the Government Research Unit in early 1987. Further boxes in this accession were found to contain some 5,000 public submissions sent in immediately following the release of the Picot Report.

Also recently available to accredited researchers by kind permission of the David Lange Estate, are six boxes of files emanating from the Prime Minister's Office concerning the Picot Report and covering the key years 1987–1988. These provide much valuable information that is available from no other source. They include correspondence with state and non-state agencies and organizations that reveal much about the way the Picot Report was viewed from within the inner circles of government. The Prime Minister's Department (now the Prime Minister's Office) ran an efficient newspaper clipping service in the immediate aftermath of the Report's release, and this material is located within the Lange papers. From a researcher's point of view, however, the most exciting find among the Lange papers were those grouped under the seemingly unexciting descriptor—Logos. On examination Logos turned out not to be about graphic design, but about design of a different description—the implementation of an ambitious public relations exercise under the joint auspices of a Wellington-based public relations firm and the Prime Minister's Department, with the express aim of selling the Picot Report's recommendations to general public.

In addition to these sources, there are also the archives of the various education boards. These furnish a useful corrective to an otherwise Wellington-centered perspective on the reforms and their impact. Much material detailing the responses of the Auckland Education Board to the Picot Commission, for instance, is held at Archives New Zealand's regional office in Auckland. Moreover, the decision to research the Picot Report as public policy presented me with the opportunity to go beyond educational records. It is important to appreciate that other major state departments displayed considerable interest in the deliberations of the Picot Taskforce, and often played a major role in the subsequent implementation of the reforms. Both the Treasury and the State Services Commission have deposited files at Archives New Zealand and today these are available for research purposes, with relatively few restrictions. Time factors and ethical constraints prevented me from conducting formal interviews with those close to the education reform process. Thanks to Graham and Susan Butterworth, however, I had access to the transcript summaries of interviews carried out by them between 1994 and 1996 as part of their Ministry of Education Oral History Project. Where appropriate, this material has been utilized as a supplement to the rich archival data now available.

Given this plethora of material, there is now little excuse for those with an interest in educational policy research not to provide broader based, historically informed, data-rich, evidenced-based accounts of educational and political change than has been possible, hitherto. Twenty years after it appeared, the Picot Report is clearly far from being finished business.

Chapter 2

Almost Alone in the World, 1942–1968

The setting up of the Picot Taskforce and the extraordinarily wide brief its members were given can only be fully understood in the context of a long-standing post–World War Two debate over New Zealand secondary education. Although this debate had its origins in an essentially interwar conflict over the aims and scope of secondary education, it was to be decisively reshaped in the postwar years by a series of reforms introduced under the first Labour government.[1]

Labour's secondary school reforms have been seen as an educational settlement whereby a new policy consensus was established after a period of crisis and struggle.[2] From the very beginning, however the establishment of universal secondary education was to be marked by a number of fundamental contradictions centering on the ultimate aims of secondary education, including equity issues, curriculum, academic standards, and school-labor market relationships. These contradictions were to remain unresolved throughout the early postwar years only to crucially reemerge in the radically changed social, cultural, economic, and policy environment of the 1970s.

Redefining Secondary Education

Secondary education for "the deserving" arrived with the Hogben-Seddon free place system in the early twentieth century. This offered free secondary education to all primary school leavers who had passed the Standard

6 Proficiency examination.³ The system's popularity, however, rapidly outgrew the projections of its authors. After World War One, the steady growth of enrolments meant that, rather than being the preserve of a sizable minority, secondary school education by the outbreak of World War Two had become a substantial majority experience. Increasing access, however, was accompanied by controversy concerning the very nature of secondary schooling. Although influential educators such as Frank Milner advocated the adoption of a common core curriculum, tensions remained over the degree of subject- and gender-based differentiation, and the prescription of a limited set of "general education" subjects.⁴

Although the new Labour government did not directly tackle secondary education until after the outbreak of World War Two, the years 1941 to 1945 have been termed "the most momentous" in New Zealand secondary education history.⁵ The school leaving age was raised to fifteen, and there were major changes to assessment processes leading to the Thomas Report, the 1945 Education (Post-Primary) Regulations, and the introduction of the "new" School Certificate.⁶ A specific secondary teacher training course was inaugurated, the secondary inspectorate doubled, careers teachers appointed, and the remaining gaps between technical and secondary schools curricula and teachers' salaries, closed.⁷

These changes to secondary education came as Labour "rejuvenated the egalitarian tradition in all its ambiguity: equality of condition and equality of opportunity for all."⁸ Over the next two decades there was virtually full employment and increasing affluence. The price was strict economic regulation and a social conformity that was increasingly at variance with global trends. Ultimately, the outcome was to be a stagnant economy, high inflation and an acute balance of payments problem. This was to necessitate an increasingly desperate political juggling act as successive governments sought to maintain affluence, temper inequalities, and ensure security through insulating the economy from overseas trends, while safeguarding a high degree of cultural uniformity.⁹

In the meantime the early postwar years provided a refurbishment of the material Utopia,¹⁰ which had shattered during the Great Depression.¹¹ Although the class conflict that had characterized much educational debate in the interwar period was de-emphasized in the early postwar years, the relationship between social groupings continued to be significant for secondary education. While the old middle class was shrinking, the new middle class was to rise steeply over the same period to reach 38.9 percent of the total. As elsewhere in the world this was accompanied by significant increases both in the new professions and in clerical occupations. In turn, this growth reflected the expansion of both state and private sector bureaucracies during the early postwar decades to incorporate some two-fifths of the workforce.[12] Expanding bureaucracies required not only specialized skills, but employees who could work efficiently within a regulated hierarchy. Labour's expanded secondary education system thus occupied a pivotal position in early postwar New Zealand, as indeed it did in the England over a broadly similar period.[13]

Raising the School Leaving Age

In 1942, director of education, C.E. Beeby intimated to the Thomas Committee that the school leaving age would be raised, making at least some post-primary schooling compulsory for virtually all pupils.[14] The move was doubtless prompted by a combination of legislative urgency and public guilt,[15] but was also influenced by wartime demands for closer links between secondary schools and industry, significant shifts in the balance of power between capital and labor, new concepts of gender roles, and the growth of school retention rates.[16] Hence, the raising of the school leaving age to fifteen was as much socially as educationally motivated.[17] Wartime youth behavior was also an issue. When in the first half of 1943 the Minister of Education, H.G.R Mason, urged his parliamentary colleagues to consider raising the school leaving age from fourteen to fifteen (and perhaps even to sixteen in the future), he emphasized that "owing to war conditions there (were) an increasing number of young adolescents who (were) missing the discipline of a normal home, and it (was) essential that the school (kept) its grip upon them during these very critical years."[18] A *Dominion* editorial in March 1943 argued that "the adolescent years of 14 to 16, a most critical period of child life, should be spent under the best and most experienced kind of school supervision in order that the development of character may receive the right emphasis and direction."[19]

The Thomas Report

With the raising of the school leaving age imminent, the broader question of how best to cater for the increasingly large numbers entering the secondary schools became urgent. Although the Thomas Committee's original terms of reference were merely to consider the choice and content of subjects for the planned new School Certificate examination, the committee went further by recommending a common core of subjects which would ensure "all post-primary pupils, irrespective of their varying occupations... a generous and well-balanced education."[20] This has subsequently been praised for taking into account "not only the intellectual, but also the moral, social and aesthetic purposes in education."[21] Certainly the committee went further than other contemporary proposals for curriculum reform such as that of Frank Milner, which was intended to apply to *secondary schools* only. A significant difference between Milner's proposal and the Thomas Committee's recommendations was that the latter expressly intended to extend their suggested compulsory common core curriculum to *all* types of post-primary institutions, which went well beyond Milner's idea of merely translating secondary schools into comprehensive institutions. Under the "Thomas" model, technical high schools, the secondary departments of district high schools, and even registered (private) secondary schools were to be included, in direct contrast to the institutional differentiation model embraced by the British Spens and Norwood Reports.[22]

From the 1980s on, however, a number of commentators have highlighted less progressive aspects of the Thomas Report: the naivety of its basic conceptions, its contradictory attitude toward the education of girls, and its essentially conservative attitudes to curriculum reform. While all children were to receive a broad, general, liberal education, girls were also to receive compulsory training in domestic science in preparation for their future role as homemakers.[23] In mathematics, differentiation between full and core courses in the report often resulted in girls being channeled into the easier of the two options, while science and technical subjects were assumed to apply only to boys.[24] Able girls took history with able boys "siphoned off into the sciences."[25] At Marlborough College, Blenheim, between 1946 and 1958, pupils were divided into three different courses: professional and general (boys and girls); trades and agriculture (boys only); commercial and home-life (girls only).[26] Thus, "the post-war woman...was to experience in her schooling, a set of cultural practices which were based on the assumptions of both a liberal ideology of equality and, at the same time, an ideology of domestic femininity."[27]

The situation for Māori was even more acute, with Māori secondary educational issues remaining essentially unaddressed. The absence of secondary education in rural areas where most Māori still lived led to the creation of the first Māori District High School in 1941.[28] John Barrington, however, has illustrated the continuing dominance of racial stereotypes that contributed to a narrow concept of the Māori future within the contemporary Education Department, citing T.A. Fletcher's, belief that, "the Māori [was] not sufficiently removed from his past to be suited for commerce." Fletcher envisaged that the core of the curriculum for the new Māori secondary departments was to be homemaking in the widest sense: for boys, building construction, furniture-making, and metalwork, and for girls, home management.[29]

Many Māori parents had for long advocated increased academic content at the expense of narrow vocational training, leading to the eventual inclusion of School Certificate courses in Māori district high schools.[30] However, Māori were far from being unanimous during this period. In February 1941 the government announced plans to expand facilities for secondary education for Māori by providing district high schools close to the centers of Māori population on the East Coast of the North Island.[31] In an open letter to the Minister of Education, R.T. Kohere, an East Coast rangatira, was highly critical of the practical curriculum it offered. Kohere was particularly concerned that the new schools did not offer matriculation as did regular high schools, thus Māori students would receive an education different from, and inferior to, their European counterparts.[32]

In response Mason defended the new schools on the grounds that they would help preserve for the Māori "his own language, culture, ideas and traditions distinct from Pakehas." The minister claimed that Māori were not being specifically disadvantaged because, "'for Pakehas, we now have technical schools instead of the older type of academic school, which was useful only to those going in for the learned professions." Hence there would be "few if any new schools of the academic type established in the near future."[33]

Other Māori complained that the government had done little to preserve Māori heritage. M. Winiata was highly critical of the "abstract concept of equality for Māori" adopted by both the Labour Party as part of its "propaganda ballyhoo," and by the National Party, based upon middle-class Pakeha prejudices, which would do away with the Māori schools and favor schools dominated by a European majority. For Winiata this concept of equality stemmed "from the idea of assimilation and baulks at anything resembling segregation."[34] Anticipating both the Te Kohanga Reo movement during the 1980s and the Picot Report, Winiata hoped that much Māori education would be marae-based, eventually evolving into something not unlike the Danish Folk High School concept.[35] Moreover, both Kohere and Winiata foreshadowed the bicultural viewpoint that increasingly underpinned criticism of secondary education offered to Māori from the mid-1970s on that was to find expression in the Picot Report and *Tomorrows' Schools*.

From School to Work

The issue of transition between school and work was another issue that would remain relatively unresolved, to be taken up more urgently in the immediate post-Picot decade. The Thomas Report downplayed the role of the school in preparing young people for work, to the extent that "education for work" became secondary to "education for life."[36] Although it rejected the British policy of discrete curricula in different types of post-primary schools, however the Report conceded that schools would provide for some differentiation in their approach.[37] Secondary schools were also encouraged to teach "vocational civics," under the aegis of social studies.[38]

Furthermore, the common core curriculum may well have provided a more efficient means of achieving differentiation rather than putting an end to it.[39] By 1944 Mason was advocating an informal system where "guidance was inseparable from pupil classification, and classification required selection."[40] The large numbers of pupils of varying abilities now entering secondary schools also meant that differentiation could be supported as one of their legitimate functions. Subsequently, they became adept at the selection and classification of pupils according to their perceived needs.[41] As Beeby himself conceded in 1958, school principals and careers advisers gave advice on courses that could amount to "strong persuasion," and there was much de facto ability grouping within the courses. Hence, despite freedom of choice, "it (was) uncommon to find a really dull child taking a straight academic course or a bright one avoiding all the more rigorous disciplines."[42] When in 1966, W.B. Sutch concluded that, "by giving the two types of schooling the one label and the one set of school buildings New Zealand did not abolish its English inheritance of treating vocational training as education and depriving the 'non-professional' children of the emphasis on the basic curriculum provided for the 'professional' children,"[43] he was revealing little that policy-makers had not already known for years.

A further issue for the future was that the more liberal tenets of the Thomas Report were destined not to be widely implemented. Pressure for examination success continued unabated.[44] The department failed to play a significant role in supporting the new curriculum.[45] Inspectors steeped in a tradition of assessing teachers for grading found it difficult to undertake the role of professional advisers.[46] Many teachers either failed to comprehend report's recommendations, or else remained opposed to them.[47] Copies of the Thomas Report often proved difficult to obtain because the original rapidly went out of print and there was no authorized reprinting until 1959.[48]

Zoning, Selection and Equity

Zoning was yet another area where crucial issues remained unresolved, only to reemerge in the immediate pre-Picot decade. In 1956 Beeby, claimed that, "In New Zealand there is no selection at all for secondary education, and, within the state system, every child, whatever his ability, is free to go to the secondary school of his parent's choice."[49] Such statements, however, were always open to serious doubt. Zoning, the defining of particular districts from which schools drew their pupils, developed as a mechanism to ensure a degree of equity at a time when schools and school populations were growing rapidly. Under its provisions, parents wishing to send their children to schools in a different zone from which they lived had to apply for permits to do so from the department.

The relationship between zoning and equity was always problematic. Zoning sometimes helped to maintain the established images and privileges of the grammar schools while perpetuating the pattern of social differentiation already evident in urban education. The Education Department was obliged during the 1950s to respond to increased pressure from parents who had secured housing in areas zoned for prestigious single-sex institutions and principals of established schools concerned about the effects of new schools on their own clientele.[50] Faced with the choice of abandoning zoning altogether, seeking enhanced legal authority to enforce existing zoning restrictions, or finding a new method of enforcing zoning with the consent of schools and parents, the Department opted to pursue the latter course.[51]

Under the new zoning system the onus of issuing permits was devolved on the principal of the out-of-zone school the parent wished the child to attend. This administratively ingenious solution made it possible to uphold widely contrasting principles. The measure of equality allegedly possessed by the original zoning concept was retained and the ideal of the neighborhood school which would provide for a cross-section of the school population within its zone was promoted. At the same time an element of freedom of choice within the state system was retained along with the introduction of an element of selectivity that permitted prestigious schools to recruit pupils from beyond their own zone.[52] These modifications to zoning allowed the Grammar Schools to retain their social prestige and academic status within the wider context of "secondary education for all" by permitting them considerable leeway to select out-of-zone pupils on the basis of academic ability. In 1955 the *PPTA Journal* noted the concerns about overcrowding expressed by school boards of "popular" post-primary schools, conceding that "many people travelled miles to avoid co-ed schools."[53] These anomalies remained, to resurface during the 1970s and 1980s.

Assessment for a New Era

Changes to examinations and to assessment procedures in secondary schools were a further aspect of the reforms that left major issues unresolved. In

1946 the first regulations for the new School Certificate Examination were gazetted.[54] The market support for this qualification was strong, and by the mid-1950s School Certificate had come to be regarded by the community as marking the successful completion of a post-primary education.[55]

Within a decade, however, School Certificate was operating in a radically different way from its inception. One unforeseen difficulty was that the secondary school retention rate rose so markedly that by the mid-1960s nearly 90 percent of all fifth formers were sitting the examination, with the result that the scarcity value of the qualification was greatly reduced. At the same time School Certificate came under attack for failing to offer a realistic goal for those pupils unable to secure an aggregate pass.[56] Although single subject passes were introduced in 1968, together with a change to the hierarchical scaling policy based on the higher quality candidates, criticism intensified. Although the examination was not actually abolished until the end of the century, the worthwhile accrediting baseline shifted upward to the sixth and seventh form levels, resulting in a credential increasingly irrelevant to the needs of secondary students.[57]

By contrast, accrediting had always been contentious. Moreover, although the elevation of the University Entrance examination into the sixth form and the introduction of accrediting theoretically freed up the first three forms within post-primary schools from the demands of the university, the long-running controversy over accrediting further illustrates the fact that not only were schools continuing to be ranked within a definite hierarchy, but that this was both widely perceived and bitterly resented.[58] Despite frequent complaints from technical schools in particular, the university and the Department of Education continued to maintain that their proper role was to avoid encroaching upon the work of the secondary schools proper.[59] As far as district high schools were concerned, the Department steadfastly maintained that such schools should concentrate on School Certificate rather than the higher qualification. This was a view which, as Howard Lee has pointed out, was at odds with Beeby's view that district high schools should perform a dual role of providing for the ambitious few as well as the majority who would not proceed to university.[60] Notwithstanding this differentiation, university authorities subsequently came under considerable pressure from the national press and many post-primary teachers to abandon accrediting altogether because of an alleged decline in standards which allowed too many students to get into tertiary education, a pressure which even withstood the release of the Parkyn Report in 1959 demonstrating that the standard of accrediting was, in fact, much more severe than the standard of external examination.[61] The seeds of popular discontent that were to lead eventually to the modular

National Certificate of Educational Achievement (NCEA) were thus well established in the early postwar years.

Defending the Reforms

Despite ongoing problems and contradictions, the architects of the secondary school reforms continued to defend them. In his 1947 Report, T.H. McCombs, the minister of education, stressed that New Zealand was in the vanguard of those countries that were trying to make democracy work, through mass education. McCombs observed that, "New Zealand and the United States (had) tried to meet the situation by giving all kinds of post-primary education, academic and practical, in the one type of school, except, in the case of New Zealand, of a few of the larger technical schools." He directly contrasted this approach with that of England and its policy of rigid selection "whereby the brightest children were 'creamed off' through competitive examination at age 11."[62] A decade later Beeby pointed out that New Zealand was almost alone in the world in having reached a position in which, "there was no selection at all for secondary education, and (where), within the state system, every child, whatever his ability, (was) free to go to the post-primary school of his parents' choice, subject only to zoning restrictions in certain areas."[63]

As radical as the New Zealand model appeared, however, it was essentially a pragmatic response to a demographic problem, "introduced, not for any doctrinaire reasons or as the results of new theories in education, but as a matter of necessity to meet a new practical situation created by the new post-primary population."[64] This situation had been brought about by the existence of two distinct issues facing postwar educational administrators: the recent rapid rise in the birthrate and the great increase in the numbers of children going on to secondary school. The first had created many difficulties for future planning because the state was obliged to prepare for record attendances in schools, while the second placed immediate pressure on the secondary schools.[65] The government's policy is best understood in relation to the fact that some 85 percent of the primary population was now going on to post-primary school.[66] Hence, the series of recent changes had occurred precisely because, "you cannot give to 85 per cent of the population the same kind of post-primary education that was originally devised for the specially selected and gifted few."[67] Thus, from the outset of the secondary education reforms educational policymakers expressed a pragmatic, even ambivalent attitude toward the purposes of secondary education that was to look increasingly threadbare in the radically changing policy environment of the 1970s.

Critiquing the Reforms

From the very beginning of the reforms, the Department was obliged to defend the secondary reforms against external criticism.[68] In 1973, the newly appointed director of education, W.L. (Bill) Renwick claimed that the controversy that surrounded the Thomas Report in the immediate postwar years had not been adequately recognized.[69] Widespread criticism about academic standards and modern teaching methods amongst employers surfaced publicly as early as October 1944, during a five day National Education Conference in Christchurch convened by the minister of education, H.G.R. Mason.[70] Many Catholics remained profoundly distrustful of both the Thomas Report and the Department's role in promoting it.[71] University academics were frequently vocal in their condemnation. From February 1942, under the pseudonym "*Grammaticus*," Professor E. Blaiklik castigated liberal education on a regular basis for more than forty years in the *Weekly News*, the *Sunday Herald*, and the *New Zealand Herald*,[72] striking a popular chord with his allegation that the new postwar curriculum had contributed to a breakdown of authority and discipline amongst youth.[73] Professor W. Anderson's pamphlet, *Flight from Reason* (1944), attracted much public sympathy with its claim that the erosion of the traditional grammar school course of core subject-disciplines had greatly weakened academic standards, and its warning that "the master of them that know, new style, is not Aristotle, but the Director of Education."[74]

A further source of critique came from the business sector. The New Zealand Chamber of Commerce set up a special Education Committee and secondary education faced considerable criticism from regional branches. In 1947 the Dunedin Chamber of Commerce convened a special meeting to which prominent local educationalists were invited. The discussion focused on the entire direction of modern post-primary education and a number of papers were presented and subsequently published in several consecutive issues of *New Zealand Commerce,* along with invited responses by selected educationalists. C.J. Wood, president of the Dunedin Chamber, laid part of the blame for an alleged decline in the basic skills of school-leavers on business hiring practices that employed qualified staff on the basic wage without much opportunity for advancement, resulting in an employee who became "stale" in his job and eventually lost incentive.[75]

However, he also issued a clear warning to educationalists as to where the real power lay:

> The education authorities in New Zealand have the monopoly in the production of our educated youth. In business we are the consumers, we are your customers, and we know what we require. If we are not satisfied, we feel that we have the right to criticise it. If the housewife goes to her grocer, or if you go to buy the goods you want, you certainly have the right of criticism, and when the producer finds that he is turning out an unsatisfactory article he makes inquiries to rectify the position.[76]

Wood's solution, that "there should be a close liaison between the schools and commerce as already we have it between the Secondary School and the university,[77] was consistent with business demands throughout the twentieth century. Similar solutions were to resurface from the early 1980s on in numerous published and unpublished Treasury and State Services Commission documents, the Sexton Report, and the Porter project. Otago Institute of Education member, A. Hanna, responded to Wood's paper by advancing the liberal-technocratic argument that the function of the post-primary school was no longer to prepare the child solely for employment. Hence, he was able to depoliticize the debate over comprehensive schooling by depicting right-wing critics as ill-informed and outdated. They recall(ed) their own school days through a haze of memory of conditions and accomplishments which in large part never existed, and then deplore(d) the fact that their child cannot spell, read, write or work simple arithmetical problems as well as they did."[78]

This early postwar debate over secondary education illustrates how educators were able to represent the new educational order as one based on scientific and modern principles. A prerequisite to any informed understanding of secondary education was held to be a specialized set of scientific and professional understandings conveyed through an increasingly specialized jargon. This attempt to give educational reform a scientific, logical basis was also a feature in other Western nations during the early postwar era, the intention being to restrict the basis of educational debate to accredited "experts."[79] Such tactics, however, could not solve the underlying dilemma of reconciling the production of a differentiated labor force for the economy with the democratic rhetoric associated with the introduction of comprehensive education.[80]

The full impact of this dilemma, though, was long delayed in New Zealand. Although the early postwar National Party displayed some residual support for prewar differentiated secondary schooling along vocational lines, it subsided when National came to power at the end of 1949.[81] The new Minister of Education, R.M. Algie frequently emphasized that his government wanted "value for money" in education, with as little wastage as possible, but he did not attempt to reverse the secondary reforms. One reason may have been that Algie saw secondary education as a form of moral trusteeship under which youth would "work in the building of their own characters, guided by capable and enthusiastic teachers," who would make them conscious of the part they would shortly have to play in the creation of a "harmonious and moral society."[82] Amongst political conservatives generally, there was also an astute recognition that comprehensive education would continue to provide the essential substance of differentiated schooling, but without its accompanying tensions.[83]

A further but still-latent difficulty for the secondary school reforms was ongoing education board resentment, centered on continuing Departmental control of schools. In November 1954, the Auckland Education Board adopted a proposal by board member Samuel S. Green, for a committee of inquiry into the administration of education in order to reduce the centralization of education.[84] Green's contention that the Department had gradually assumed powers and functions that it was never intended to have was a further articulation of the nation's regional educational authorities historical concern that centralization had gone too far.[85]

During the early postwar years, however, criticism of secondary schools was deflected by a growing cadre of academics who owed their positions to the postwar expansion of teachers' colleges and university-based education Departments. In February 1956, a series of articles in the *New Zealand Herald* blamed poor handwriting and vandalism on the fact that teachers were no longer concerned with character building. This prompted a caustic reply to the editor from all seven education department staff at Auckland University College protesting the paper's derogatory editorials on education, and complaining that to base, "general attacks on unverified opinions and particular incidents, is, as you must realize, illogical and banal."[86] The Department's relief at this support was revealed by L.F. le. Ensor, the Auckland District Superintendent of Education. Ensor sent Beeby the full text of the Auckland University Education Department response, astutely observing that; "We have some friends in Auckland who are nothing if not direct, and plain spoken."[87] Despite the appearance of these invaluable allies, however, criticism of secondary schooling and of a Department viewed as the lynchpin of support for the reforms was never far from the surface.[88] Particularly from the late 1970s on, major shifts in the economic, political, and intellectual environment were to result in many erstwhile friends joining the ranks of the critics.

As we have seen, the early postwar secondary schools were always regarded in part as an investment in human potential and hence, future economic productivity.[89] One aspect of this was the encouragement given to curriculum innovations such as the new mathematics from the late 1950s on, which were believed to lead directly to the creation of a technologically sophisticated work force able to fuel economic transformation.[90] These links between education and efficiency sharply increased during the 1960s, but rapidly became a double-edged sword for educators. Some indication of the problems ahead for secondary education can be seen in three key reports: the Parry Report (1959), *School and Nation* (1961), and the Currie Report (1962).

The Parry Report

Although the Parry Report focused on universities rather than secondary schools, its assumptions and conclusions were to be crucial in signaling a further and ultimately decisive drift toward economic and social efficiency. The Parry Committee's terms of reference were to "indicate ways in which the university system should be organized to ensure that the long-term pattern of developments is in the best interests of the nation."[91] Professor F.W. Holmes, McCarthy professor of economics at Victoria University, and Dr. W.B. Sutch, permanent secretary for Industries and Commerce, were key influences. During the 1950s Sutch in particular promoted a concept of development based on rapid industrialization underpinned by a skilled workforce.[92] By 1957–1958 a growing economic crisis threatened the Labour government's policy of full employment leading both Sutch and Holmes to express alarm concerning New Zealand's economic future with Britain's expected entry into the European Community. Anticipating the concerns of commentators such as S. Harvey Franklin some two decades later, the Parry Committee warned that if the country's standard of living were to continue to improve, the volume of its production would have to increase more rapidly than its population.[93] To accomplish this, greater emphasis was to be placed on skill and efficiency in the workplace, thus investment in education was) "at least as vital as adequate investment in physical capital."[94]

The report also laid particular emphasis on the demand and supply of university graduates, with a strong focus on the efficient production of scientists, technologists, and business managers, with the least possible wastage.[95] Estimates of future needs would be "related to the economic, social and cultural objectives of society."[96] Here indeed lay an ominous message for secondary as well as university educators had they cared to note it, for if educational developments designed primarily to safeguard future economic and social stability were not forthcoming, then the state was clearly reserving the right to use its power to direct where appropriate and to prune when necessary.

Contemporary educators, however, misread the message. Shortly after the release of the Parry Report, a *PPTA Journal* editorial correctly observed that it had called for a revaluation of education, but took this to mean that expenditure on education should be increased on the grounds that "if we are ever to have anything more than a cowshed economy and culture we must have first-rate universities and schools."[97] Public opinion appears to have been far less convinced that more expenditure would bring about appropriate results. McLaren observes that during this period public discussion of education was highly emotive, typified by an *Evening Post* editorial in April 1958 claiming that the New Zealand school system was one in which children learned "the least possible in the longest possible time."[98]

Growing public unease over secondary education's role in shaping the nation's economic future was sharpened by the fact that there had been rapid growth in educational expenditure at a time when government expenditure in general had remained relatively constant.[99] Between 1950 and 1960 educational expenditure had increased nearly three-fold, while overall government spending had risen only marginally.[100] Public concern over educational standards was linked to the apparent failure of secondary education to deal effectively with youth problems, a factor which had been an influence on the wartime raising of the school leaving age. Juvenile delinquency was by 1960, a significant national issue.[101] In 1954 the Mazengarb Report, copies of which were distributed free to every New Zealand household, had called for urgent action on the problem.[102] In September 1960, when the Currie Committee was still in the early stages of its deliberations, the widely published disruption of the annual Hastings Blossom Festival by rampaging youths focused further attention on the shortcomings of secondary education. A cartoon in a major daily newspaper depicted lax school discipline and mediocre secondary school academic standards as being responsible for the debacle, and even the ensuing debate amongst educators focused on how secondary schools alienated mediocre students while failing to challenge the able.[103]

School and Nation

During this period Left-liberal educators were increasingly critical of a secondary system they saw as conservative and outmoded. The American sabbatical visitor, Professor D.P. Ausubel told a conference of Department of Justice psychologists in 1958 that "the secondary system impressed the overseas visitor as the most anachronistic segment of New Zealand educational life."[104] In September 1960 Phoebe Meikle, a highly regarded educational commentator and teacher,[105] published a critical essay on secondary education subsequently reissued as the pamphlet, *School and Nation*, in July 1961 by the New Zealand Council for Educational Research. Meikle accepted the broad philosophy of the Thomas Report, but was highly critical of the way many schools had subsequently interpreted its recommendations.[106] Meikle claimed that the secondary schools system failed the majority of its pupils, failing to challenge the able students, and alienating the less able, leading to serious behavior problems.[107] Even the large group in the middle were "victims of two paradoxes of our egalitarian society: our selection of a pass in the School Certificate examination after three year's post-primary schooling as the symbol of intellectual 'normality'; and the prestige attached to particular courses and subjects."[108] The result of all this had been an actual decline in academic standards, particularly in English, as well as considerable waste of human resources, with consequent dissatisfaction among pupils, teachers, and parents. Thus for Meikle

> the strengths and weaknesses of a system of education in which state-controlled schools staffed by state-educated, state-trained teachers cater for almost all a nation's children, must be those of the nation. That is an inescapable fact.[109]

Meikle's perceptive pamphlet laid bare the basic contradictions in postwar secondary education. The educational bureaucracy, however, promptly closed ranks and Meikle, profoundly discouraged, left teaching for a career in publishing. The issues she raised were to endure.

The Currie Report

The Currie Commission was set up in February 1960. Over the next twenty-eight months it heard oral testimony from around the country and received over 400 public submissions, releasing its final report in July 1962.[110] David Scott has demonstrated the extent to which the Currie Commissioners were "captured" by the New Zealand Department of Education.[111] Accordingly, their report began by reasserting Fraser's 1939 objective of equality of educational opportunity.[112] The Report was also highly selective in its utilization of public submissions. It downplayed continuing doubts about secondary education expressed from commercial and industrial quarters such as the Associated Chambers of Commerce submission, that reiterated its 1947 position on secondary education and advocated a more restricted form of post-primary education that would have seen the reintroduction of a selective examination at the end of the second post-primary year in order to channel students into "appropriate" vocations.[113] In similar vein a submission from the combined Engineering and Metal Trades claimed that the cultural emphasis of education had, "failed miserably, tending to produce misfits ill-equipped to earn a living rather than useful citizens," and recommending that the chief aim of education should be vocational rather than cultural.[114]

The Department of Education, however, in a series of well-presented submissions to the Currie Commission expressed unreserved support for the comprehensive ideal. Noting that, "without any exception, this is the kind of school which has been established in New Zealand in the period of rapid growth since the end of World War II," the Department asserted that multi-course schools had achieved considerable success with a wide range of pupils, even in comparison with supposedly more academic schools.[115] It went on to warn that, if such schools were to be neglected in the future, it would, "result in the abandonment of the basic idea of the community school and would bring with it inevitable problems of selection of pupils."[116]

Perhaps not surprisingly, given its reliance on the Department of Education for much of its information, travel arrangements, itinerary and secretarial assistance, the Currie Commission recommended against any major overhaul of the education system.[117] In adhering to this view, it could count on two major factors. The majority of educational professionals, while sometimes critical of secondary education in practice, were not prepared to totally abandon the 1940s reforms. For instance, J.H. Murdoch, the author of *The High Schools of New Zealand*, expressed general support for the notion of a common core of subjects, but conceded that, "its compulsory application to all and sundry, now that the general position has been gained, surely calls for re-examination... (now that)... the danger of a counter-revolution (had) largely disappeared."[118] Furthermore, in the early 1960s secondary schooling could still derive some benefit from the conviction that increased educational expenditure could be justified in terms of its economic payoff. Sutch, whose views had strongly influenced the earlier Parry Report, articulated this view in his lengthy submission to the commission. Entitled "Education for New Zealand's Future," this warned that "failure to put more resources now into education will lead very soon to a considerable reduction in the rate of economic growth and probably to an absolute fall in levels of consumption per person."[119]

Although the commission could safely reject demands for a return to prewar selective vocationalism in secondary education, it could not ignore the growing political conviction that secondary schools should play a major role in producing workers and citizens. The solution seemed to lie in asserting that "in a world that has suddenly become more competitive and where a very different political and economic system is challenging democratic ideas, education was vital to nations wishing to maintain or improve their positions." As Sutch explained to the Commission in his submission, New Zealand was

> faced with an economic challenge which is peculiar to this country at its present phase of development [which] arrives at the same conclusion—that we must make the best use of the abilities of the whole population and that one of the keys to this lies with education. Equality and expediency appear, therefore, to point in the same direction.[120]

Undoubtedly, the commission subscribed to the contemporary notion that the entire population constituted the natural wealth of any nation, thus not to educate them to their maximum capacity left a significant part of the nation's natural resources undeveloped.[121] This view had parallels elsewhere. In the United Kingdom during the same period, documents such as the Crowther Report (1959) can be seen as an attempt to reconcile a traditional socialist commitment to educational equalization, with a plausible analysis of the requirements of postwar capital.[122] Central to this viewpoint was a particular view of the relationship between schooling and employment.[123] The Currie commissioners embraced economic challenges of the future through (following the contemporary United States Rockefeller Report), "the pursuit of excellence" in education. In addition the Commission noted several points from the Sutch submission:

1. That both manufacturing and servicing industries must expand considerably if production was to rise more rapidly than population and yield higher living standards.
2. The possibility that future development might have to be undertaken in an atmosphere of less favorable terms for overseas trade.
3. The apprehension aroused by relatively slow economic growth in comparison with other countries, indicating inefficient use of resources and a need to pay closer attention to the role of education in "...developing skills of all kinds and the arts of management."
4. The need for economic flexibility allowing for transferability of the labor force "...to more essential and more profitable employment."

The contribution of education to raising levels of skill and knowledge in under-developed areas.[124]

As we shall see in part 2 of this book these views were to be reiterated even more emphatically from the 1970s on, finding their ultimate expression in the Porter Report in the early 1990s. In an era of expansion though, the commission was able to use such arguments to reject calls for early vocational specialization and rigid controls on the numbers proceeding on through the post-primary education system in favor of a set of more ambitious recommendations. These included raising secondary academic standards through increased teacher training and funding, the revision of the School Certificate examination to meet the diverse needs of pupils and employers, a greater provision in the curriculum for non-academic children, and the eventual raising of the school leaving age to 16.[125]

In general, secondary teachers had reason to be satisfied with the commission's findings. The *PPTA Journal* probably spoke for the majority of its members when it commented shortly after the Report's public release in July 1962, that "there can be no dispute that in two main particulars, the commission was dead on target with its stress on the revaluation of the teaching service and the urgent need to improve teacher training."[126]

The long-term cost, however, was a failure to address what S. Harvey Franklin was later to identify as a growing economic and cultural crisis that, by the early 1980s, could no longer be ignored. McCulloch has pointed out that the Currie Report tended to ignore or marginalize groups that did not fit in with its theories of progress.[127] In this respect, it was not alone. Commenting on contemporary British education reports such as the Crowther Report (1959) and the Newsom Report (1963), the authors of *Unpopular Education* claimed that there was no apparent discomfort about gender among policy-makers at this time, with the sexes being accepted unquestioningly as having different roles and needs which were to be reflected in the curriculum.[128] In New Zealand, the highly conservative views of John Newsom on the education of British girls, while not directly affecting the official curriculum in New Zealand, were evident in the popular press and in some educational writing.[129] Thus, a considerable ambivalence over the purposes of secondary education for girls among policy-makers remained to fuel further concerns from the mid-1970s on.

Although it rejected the Thomas Report's view of gender specialization, this was not so much an indication of advancing social liberalism, as it was a recognition that changing conditions had "resulted in the employment of women to a degree that could not have been imagined 30 years ago." Although the Commission recommended that technical training be extended so that the country could make "greater use of the potential skill and ability of its women,"[130] it clearly regarded young women "as a pool of untapped talent, as useful to the country's future growth, rather than as a central mechanism for redressing gender-based inequalities and enhancing social equality."[131] Similar views underpinned the Science and Mathematics New Scheme introduced in 1957 in order to fast-track female school leavers into basic science and mathematics courses, leading to a career in secondary teaching.[132]

For Māori too, the Currie Report was to reflect the dominant European view. The *Report on Department of Māori Affairs*, 1960 (the Hunn Report), which was to become the basis of government attitudes toward Māori throughout the subsequent decade,[133] had already set much of the tone for the Currie Commissioners in its redefinition of official policy as integration rather than assimilation.[134] The Hunn Report saw Māori education as "the one thing more than any other that will pave the way top further progress in housing, health, employment and acculturation."[135] Hence, it recommended the setting up of the Māori Education foundation to finance secondary and university education for Māori students through a system of competitive scholarships, a recommendation hailed by one contemporary research paper as "the most dramatic and effective suggestion to stem from the Report."[136] Māori opinion, however, was less complimentary. A booklet released shortly after the publication of the Hunn Report by the Māori Synod of the Presbyterian Church warned that the term "integration" embodied an assumption "that we must forget our history, our culture, our racial origins—all that is included in our Māoritanga."[137] Writing much later, Walker concluded that the authors of the report failed to take into account the process of enculturation in the development of identity,[138] In similar vein Graham Hingangaroa Smith and Linda Tuhiwai Smith have pointed out that as far as education was concerned, Māori children were being perceived as being culturally different in their own land.[139]

The Currie Report shared the Hunn Report's view of Māori education as a problem which required special attention, though it conceded that Māori opinion had been inadequately canvassed.[140] It identified two educational problems which required "solving": the first being to determine how far the education offered to Māori was suited to Māori needs; and the second, to discover how best to enable Māori to take advantage of the education being offered.[141] Admitting that previous policies that sought to adapt the Māori to a European type of education had failed, the Report advocated that "such elements of his Māori background must be included in his schooling as will given him still the sense of belonging to a race of known and respected culture."[142] At the same time it viewed cultural impoverishment and low socioeconomic status as interrelated, given that "too many live in large families, in inadequately sized and even primitive homes, lacking privacy, quiet and even light for study, and too often there is a dearth of books, pictures, educative material generally, to stimulate the growing child."[143]

For these reasons the Commissioners believed that Māori education "must become an area of special need, requiring special measures and, inevitably, increased expenditure."[144] Turning to secondary education, they recommended that schools, including district high schools and secondary schools, be defined "Māori service schools" wherever they contained a substantial proportion of Māori students.[145] They also encouraged more Māori participation in local educational administration,[146] an expansion of the Māori intake into teachers' colleges,[147] and an extension of vocational guidance and parent education,[148] but rejected any special adaptation of School certificate English for Māori pupils as "a false kindness."[149]

As was the case with female students, Māori secondary students were principally viewed as an untapped national resource, potentially able, through enhanced educational opportunities to make a significant contribution to the national welfare. Such views were made even more explicit in a paper presented by Sutch to Māori students at Victoria University, Wellington, in May 1964.[150] Sutch warned his audience that until the Māori "caught up" educationally, they would be unable to "make the economic contribution to the future that the country need(ed),"[151] For this reason he wanted to retain Māori children in school for longer periods of time, even if this meant raising the school leaving age to sixteen.

As we have seen, along with ambivalent attitudes toward both women and Māori, contemporary educational commentary often failed to appreciate that, by attempting to sell education as an investment strategy for future economic expansion they were sowing the seeds for future difficulties when times became harder. In the United Kingdom during the same period, social democratic philosophies faltered precisely because they failed to achieve economic growth and order.[152] In New Zealand, even more ominous developments for secondary education lay on the horizon. Malcolm McKinnon has suggested that one response to slowing economic growth was the arrival in Treasury during the 1960s of a new generation of recruits who were high-achieving economics graduates rather than the traditional civil servants who began careers as public service cadets.[153] The result was to be a dramatic transformation that saw Treasury in the next decade increasingly emphasizing both its role as the main advisory body on economic policy, and the necessity of short and medium term economic management.[154] Moreover, as in the United Kingdom slightly earlier in the decade, a relatively low rate of economic growth fostered official enthusiasm for "indicative" planning, a move actively facilitated by the long-standing friendship between Sir Frank Holmes and the new treasury secretary, Henry Lang. The National Development Conference (NDC) in 1968, convened in response to the 1966/1967 balance of payments crisis, was an early example of such planning.[155]

As far as secondary education was concerned the Parry Report had issued a clear message that state resources were not only limited but might be justifiably switched to areas of further training identified by future governments as constituting national priorities. By the end of the 1960s, with educational expenditure under increasing scrutiny, the minister of finance, R.D. Muldoon was warning that, given limited resources, there was a "strong argument for some bias in application of resources to those areas which actually assist[ed] New Zealand economic development."[156] Moreover, by this time it was increasingly common to find secondary education in particular being singled out as the focal point of an emerging global "educational crisis." Though they differed as to the precise nature of the crisis, commentators shared a common view that expansionary education policies had come to an end and that continuously rising educational expenditure could no longer be sustained in the face of both rising demand and dwindling resources. Phillip Coombs' influential book, *The World Educational Crisis* (1968), for instance, claimed that there were two aspects to the international educational crisis: a great increase in demand for education caused by the rising birthrate, multiplied by increasing educational aspirations; and the "dwindling resources available for the educational cuckoo whose demands are apparently insatiable and are everywhere rising at a rate much greater that national income." More ominously for the future perhaps, Coombes saw the solution to the crisis, not in terms of further educational expansion, but in a reexamination of the efficiency of existing systems.

In New Zealand similar misgivings were increasingly evident at the end of the decade. Writing in *Delta,* P.J. Mundy warned that as additional funding for secondary education was unlikely, existing resources would have to be deployed more efficiently.[157] *Delta* editor Richard Bates was critical of those he labeled, "defenders of the faith," seeing in the recently published, *New Zealand Education Today* (1968), by well-known educator F.W. Mitchell, "a tacit refusal throughout... to deal openly with the contemporary crisis."[158] These were to be prophetic words. As the next part of the book will demonstrate, the next two decades were to see the culmination of a growing economic and cultural crisis in New Zealand that was to impact directly on secondary education.

Part 2

Crises and Solutions

Chapter 3

Game War

Although early postwar secondary education faced increasing tensions and contradictions, the tenor of the debate that preceded the setting up of the Picot Taskforce can only be fully appreciated in the context of the profound New Zealand economic and cultural crisis that deeply influenced, and eventually came to dominate, the ongoing debate over secondary education during the 1970s and early 1980s. Indeed, when the preoccupations of the taskforce and the recommendations of their Report are viewed as part of a wider policy response to deep-seated structural problems that were widely perceived to have worsened, they become far more understandable. Because so many events and changes were occurring simultaneously during this period, each of the three chapters in this part of the book concentrates on particular themes. This, the initial chapter in part 2, opens by outlining the growing economic and cultural foment in New Zealand, especially as this was interpreted by some of leading contemporary commentators. It then turns to a description and analysis of some influential sector/group responses.

Radical Problems

Malcolm McKinnon has shown that, in response to the general economic slowdown that began during the 1960s, the next decade was to witness "the beginning of a seismic shift from trust in governments to a trust in markets—a reflection, it could be said, of a reverse movement of the Great Depression era."[1] In 1978 the Bank of Canada came to embrace monetarism. In the following year Margaret Thatcher came to power in Britain and Ronald Reagan became president of the United States in January 1981. Closer to home in Australia, Treasury thinking during the 1970s reflected the international breakdown in the hitherto optimistic social meliorist view of human nature that was central to Keynesian model. Globally, this shift had already been foreshadowed by decisive shifts in World Bank, International Monetary Fund and OECD thinking, evidenced by their publications.[2]

The situation was, however, considerably more acute in New Zealand. As early as the second half of the 1960s mounting economic difficulties threatened national prosperity. There was a movement toward corporate ownership concentrated within a few large multinational companies, although the growth of overseas ownership over a widening area of the economy steadily gained momentum.[3] These changes were accompanied by significant shifts in income distribution, with the range between high and low incomes steadily increasing.[4]

Successive governments failed to produce policies to deal effectively with rapidly changing social and economic conditions. The country as a whole continued to pursue a Third World commodity export strategy until well into the postwar period. Although in the short-term this strategy delivered high living standards, it eventually suffered from price volatility and vulnerability to foreign competition.[5] Markets and prices came under increasing pressure in the postwar era, exacerbated by the increasing unwillingness of the United Kingdom, the country's chief export market, to accept unlimited agricultural imports, the postwar recovery of Western Europe, and the creation of the European community. These changes had an increasingly severe impact on the New Zealand economy, with the first real crisis occurring in 1966, when the commodity terms of trade fell sharply. Except for a misleading recovery in the 1972–1973 period export prices were to remain depressed.[6] Although New Zealand increased exports during the 1970s, this strategy had only limited success, largely because expenditure on imports grew at an even higher rate. Particularly after 1975 the legacy was to be a massive fiscal deficit, financed by heavy overseas borrowing.[7]

From an early twenty-first century perspective, New Zealand in the second half of the twentieth century exemplifies a classic example of "riches-to-rags." In 1945, when the last of the post-primary reforms were being implemented, the country was on a per capita basis, one of the world's wealthiest nations. By 1993, nearly a decade *after* the radical reforms set in place to improve economic performance by the Labour government, it had slipped to twenty-third with a GNP per capita of US$12,060. Moreover, the actual ranking disguised how far the country had slipped in relative terms to other Western nations such as Switzerland

(US$23,120), Canada (seventeenth, US$20,320), Australia (nineteenth; US$17,070), or the United Kingdom (eighteenth; US$17,760).[8]

Radical Solutions

Perhaps to an even greater extent than elsewhere, New Zealand was to experience volatile shifts in both its intellectual climate and its policy environment during these years. In addition to the changes in Treasury outlined in the previous chapter, the academic departments of economics in New Zealand universities were becoming more market-orientated. The Planning Council under Holmes published *The Welfare State: Social Policy in the 1980s*, a policy document that distinctly foreshadowed arguments over the next decade centering on the fiscal burden and on disincentives associated with current fiscal policies.[9]

With successive governments failing to act decisively, worsening socioeconomic problems increasingly became the subject of discord, leading to radical proposals for change. Writing in 1978, Victoria University professor of geography, S. Harvey Franklin characterized New Zealand as a dependent society that was both anxious and contradictory.[10] In a book that was to have a significant political impact, Franklin employed an analysis that was to clearly anticipate those of Lange's Third Labour government, Treasury, and the State Services Commission (SSC), post-1984. Franklin warned that successive governments had become the hostages of vested interests created by the welfare state. He characterized New Zealand's society and economy as the product of a consistent desire to modify the worst features of capitalism and international trade through a mixture of welfarism and national socialism.[11] Since the 1967 recession there had been a very low rate of increase in outputs per head of labor, subsequent economic growth occurring only because the steady demand for rising personal and public consumption had been satisfied by successive governments with little regard to the fluctuations that had occurred in export earnings. Hence

> there is a repetitive quality to public life that in the end is disenchanting and productive of ennui and cynicism. The society passes from one crisis to the next, from one confrontation between labour, management and [government], from one dispute and counter-argument to another. Inevitably each party attacks the other and a game war emerges where the settlement of old scores appears to take precedence over the search for agreement. Bystanders inevitably attribute blame, as do the participants. Inflation would halt if unions reduced or abandoned their demands; things would be better if government left private

enterprises alone; or better if manufacturers were provided with less protection; or if there were fewer university students.[12]

As Franklin saw it, New Zealand faced two related groups of structural problems. The first group centered on the urgent need for radical economic restructuring. The welfare state was disproportionate to investment. The solution lay in opening up the economy, even at the price of a government courting electoral defeat.[13] The second group of problems was essentially cultural. Franklin was particularly critical of the New Zealand secondary school curriculum that, he contended, had helped to create the cultural mentality of the welfare state. Aware that the Educational Development Conference's recently published Wording Party report on educational aims and objectives had emphasized the moral purpose of education, Franklin considered such sentiments to be noble in intent, but an inadequate foundation in the absence of a suitable economic base.[14] Both these groups of structural problems identified by Franklin; the economic and the cultural, were to have a significant impact not only on the challenges secondary education and the Department of Education now faced, but also on the way they were to be increasingly viewed by a new generation of educational critics.

Significantly, both the problem and possible solutions to the crisis identified by Franklin were being addressed at much the same time in a 1978 discussion paper prepared for the New Zealand Planning Council by a special working group headed by Brian Picot.[15] Picot, a social liberal who was to later speak out against apartheid in South Africa,[16] observed that the group had reached three important conclusions—that human relationships were complex and would change only gradually; that conflict was inherent in present circumstances, requiring effective conflict resolution, and that next decade would see workplace changes reflecting increased worker participation and teamwork. The first section of the paper, provocatively entitled "An Economy Under Siege," warned that present trends threatened economic well-being as much as had the Great Depression. The value of exports had declined and heavy government borrowing had created a huge national debt, accompanied by high inflation. Within society, unrealistic and adversarial attitudes prevailed, "a destructive conflict that undermines our struggle for recovery."[17]

Picot believed, however, that there was a broad commitment to a mixed economy. Hence he saw the solution as industrial democracy, with worker and management working closely together, "as happens on company boards in Europe."[18] Such a solution was indeed urgent because we can "either find ways of working together effectively for progress or, as a nation, become increasingly divided, unstable and poor."[19]

In a privately published address to the Auckland Chamber of Commerce in early 1979 Picot reemphasized that New Zealand faced an uncertain economic future. Economic policies alone were insufficient because it would be, "our attitudes that will determine whether or not New Zealanders will halt the present drift into second class citizens of the world."[20] This conclusion led Picot to an embrace a more inclusive concept of community that would later infuse the Picot Report. He believed that economic woes had coincided with a similar deterioration in community attitudes marked by "the unintelligent protection of personal and tribal interests—and to hell with the other fellow."[21] Comparing the failure of the British model of industrial relations that foreshadowed disaster for its participants with the success of the West German collaborative approach to national development, Picot argued that all sectors of the New Zealand community had to unite to solve the country's problems.[22]

Secondary School Radicalism

Secondary teachers responded in several ways to the unfolding economic, cultural and educational crisis. One response was a rapid increase in teacher militancy with the result that the Post-Primary Teachers' Association (PPTA) had, by the late 1960s become a new-look professional organization, able not only to focus on problems teachers had in common, but also more willing to organize mass campaigns aimed at bringing the problems of secondary schools to the attention of the politicians and the public at large.

This new approach was to have significant consequences for secondary education. Writing at the close of the 1970s, education historian John Barrington contended that, "Recent increases in teacher militancy, bringing with it conflict with the Director-General of Education and the Department, demonstrates how, particularly at the secondary level, there can be a fragility about the existing balance of interests in school government between the central authority, national teachers organizations and local secondary school boards of governors."[23] Increasing militancy was to further feed public perceptions that, not only were secondary schools in crisis, but that the interests of parents and students were essentially different from those of teachers. The result was that, by the 1980s education in general and secondary education in particular had largely lost the battle with public perception.

The portrayal of schools and teachers in the nation's daily newspapers was another factor in the negative public perception of secondary education. Dorothy Roulston's examination of five major daily newspapers in the 1978–1982 period highlights their significance as gatekeepers in the selection and presentation of educational news.[24] She concluded that the newspaper-reading public during this period displayed a greater awareness of educational issues and that educational reporting widened to include issues such as unemployment, learning difficulties and youth behavior.[25] Moreover, a new style of educational reporting appears to have emerged from the mid-1970s as secondary education became the site of more frequent conflict. One example of this changed style is the reporting of planned industrial action by secondary teachers in early 1978. The lead-up to that strike, as reported in two newspapers, the *Evening Standard* (Palmerston North) and the *Christchurch Star*, is particularly illuminating.

As early as August 1977 the *Christchurch Star* reported that attitudes to educational spending had come into sharp conflict at the PPTA's annual conference in Wellington, with an exchange between the Prime Minister R.D. Muldoon and the president of the PPTA, Mrs. Ida Gaskin.[26] In November came reports of a series of PPTA regional meetings over salary negotiations. Following the one-day strike of February 23, 1978, there was particular emphasis in the newspapers on those teachers who, despite the pressures placed upon them, had remained on the job. Minister of education, Les Gandar was quoted as having praised the professionalism of the teachers who would not strike, and depicted the children as victims.[27] By March 1978, the Auckland region of the PPTA was reported as requesting the PPTA Executive to organize rolling strikes throughout the country which would continue until existing salary claims were settled.

Throughout the comparatively brief time when strikes were big news, educational reporting was characterized by a concentration on isolated images rather than on sequential analysis.[28] "Putative deviation" was assigned from which further myth-making and labeling could then proceed.[29] The aberrant and the abnormal became the norm in educational reporting. Thus teachers *protested* over salary claims;[30] they *claimed discrimination*;[31] they were *defiant*;[32] they *slated* the government;[33] they *argued*.[34] The schools themselves were depicted as sites where students were threatened by sex, violence and drugs,[35] and where vandalism occurred on a relatively regular basis, rather than as learning institutions.[36] Moreover, teachers were perceived in a number of newspaper stories to be teaching from controversial materials such as the United States-developed social studies program, MACOS, as well as remaining committed to liberal teaching methods despite these having been discredited by overseas educational "experts."[37]

Although PPTA was often depicted in the media as a radical union bent on wrecking secondary education, the union's growing interest in curriculum and assessment matters perhaps better illustrates a growing perception amongst secondary teachers that education was essentially political in nature. Embracing what was to become known as the "professional project" led PPTA to focus on the content and control of the curriculum by teachers as well as on assessment, leading to an increasingly politicized debate over educational outcomes.[38] The 1965 Annual Conference expressed concerns about the inability of School Certificate to meet the needs of all secondary students leading in the following year to a call for PPTA to adopt a clear philosophy in respect to qualifications.[39] PPTA subsequently initiated the first comprehensive examination of the secondary school system since the Thomas Report, adopting a broad definition of the school curriculum to encompass virtually all school activities, including assessment. In some respects, this development brought PPTA and the Department into increasingly close collaboration. The result was a series of booklets on the secondary school curriculum that were widely distributed in secondary schools.[40] The first of these, *Education in Change* (1969), was published by the PPTA's Curriculum Review Group.[41] Judie Alison has recently pointed to the influence of the contemporary "objectives movement" in the United States and particularly writers such as B.S. Bloom and R.W. Tyler, whereby objectives were explained as being "performance-orientated." Alison argues that such language anticipated later discourses around the New Zealand Qualifications Framework (NZQF).[42] *Education in Change* also went far beyond the Thomas Report in endorsing a life-adjustment and social meliorist approach to curriculum, not only placing emphasis on student enquiry, contemporary society, and self-respect, but also linking both curriculum and assessment to changes in the wider community, including those taking place in the economy:

- the development of a unified, cooperative world society.
- the need for new skills and products to enable New Zealand to compete in world markets. the need for cultural understanding of Māori and Pacific Islands students
- the changing structure of the family including its declining ability to teach values.
- a weakening influence of churches
- the more assertive role of women
- changes in adolescent behavior
- the impact of the mass media
- the growth of science and technology leading to the need for new forms of enquiry

- changes in the secondary system including increased retention rates resulting in a dramatic expansion of the school's role and responsibilities.[43]

The third book in the series, published in 1972, saw educational change as vital because New Zealand society was in a state of change and tension, leading to a lack of shared goals and hence, to considerable ambiguity in the role of the secondary school.[44] Future reform should therefore recognize "the school as a unit, with its approved curriculum based on its own needs, and evolved by its own staff."[45]

The appearance of these publications clearly signaled a shift in emphasis from equality of opportunity, to educational outcomes[46]. They also fostered further public suspicion that the Department of Education had become closely aligned with PPTA. In an attempt to forestall charges of provider capture, the Director of Education Dr K.J. Sheen, felt it necessary to reassure National's Minister of Education that his Department had

> never construed it as its duty to lead public agitation for educational changes—a role that teacher's associations would like to thrust on it. It has nevertheless considered it its duty to initiate change with the agreement of Government and as far as possible in association with the teachers' organizations.[47]

In many respects, however, Department and PPTA were in agreement over the broad direction of future secondary curriculum reforms. Between August 1971 and September 1973 the Department conducted three Lopdell House Courses involving principals, secondary teachers, teachers college staff and departmental officers, aimed at fostering new ideas about curriculum and assessment. Both Department and PPTA drew on a wave of current American educational liberalism that castigated the school system as a soulless bureaucracy that produced failure, including Postman and Weingarter's *Teaching as a Subversive Activity*; Charles Silberman's, *Crisis in the Classroom* (1970), and William Glasser's, *Schools without Failure* (1969). At the same time, a PPTA submission to the 1972 Lopdell House Conference again advocated the clarification of educational objectives, referring to specifically to a set of minutely defined sets of objectives favored by Bloom.[48] Clearly at this stage then, the notion of "schools without failure" was by no means regarded as being incompatible with outcomes based approaches to assessment.

The election of Labour in 1972, however, after more than a decade of relatively conservative and cautious National governments, was to heighten the Department's dilemma in trying to accommodate the PPTA while portraying itself as an honest broker between union demands and public opinion. The new Director of Education, A.N. Dobbs, a career civil servant rather than an educationalist in the Beeby mould, incurred the new Labour government's particular hostility. Ironically, after ensuring two decades of complaints that it had moved too quickly into a new era, the Department was now accused of being hidebound and resistant to reform.[49] Animosity spilled over into the public arena, as when in early 1975 the Minister of Education, Phillip Amos, allegedly told a group of teacher trainees that it would take a decade "to remove all the Tory deadwood cluttering up his Department."[50] Accordingly, Amos influenced the appointment of new senior officials within the Department whom Labour regarded as being sympathetic to curriculum reform. Following Dobb's retirement, Renwick was appointed director of education, and Peter Boag, previously PPTA general secretary, became director of secondary education.[51]

These key political appointments occurred at the very time when forces within the PPTA were pressing for further curriculum reforms involving more radical constructivist goals for the curriculum. The impact was to be twofold. First, continuing militancy brought the PPTA into further conflict with the politicians. Although, as in Australia, post-primary teachers increasingly saw themselves as workers, the tension between the roles of worker and professional grew as confrontation with successive governments over salary claims accentuated conflicts between "moderates" and "radicals" over the place of direct action.[52] Second, the nature of the PPTA Executive continued to change as the so-called "Young Turks," a younger, more radical and more-union orientated group gained power. In turn these new developments were to be influenced by the rise of radical new advocacy groups with a strong base in university education faculties during the 1970s.

The Rise of Radical Advocacy

The growing economic and cultural crisis was to have a signal impact on the growth of radical advocacy directly largely at secondary education structures and outcomes. The crucial difference between the radical advocacy groups of the 1970s and early 1980s and earlier educational critics was that the former emphatically rejected the Enlightenment universalism that had underpinned earlier liberal educational reforms, emphasizing instead a blend of cultural relativism and ethnic particularism known as cultural essentialism, or culturalism.[53] As in other Western countries such as the United Kingdom and the United States, the ambivalent position of New

Zealand's own Liberal-Left as a relatively privileged new middle class of "caring" professionals was largely responsible for its retreat from class to culturalist politics throughout the Western world.[54] From the late 1960s onward many middle-class radicals trained in the humanities and social sciences now found acceptable sanctuaries in the welfare and creative professions, thus benefiting from capitalism's operation while at the same time avoiding direct involvement in capitalist enterprises. It was a way to "exercise their talents without compromising their radical political ideals."[55]

With its liberal guilt hoisted on the uncomfortable petard of what Alvin Gouldner termed a "goodness and power" paradox, the radicalized new middle class of the 1970s and 1980s underwent a decisive shift from class to cultural politics.[56] According to the new understandings of identity politics, it was no longer the proletariat that experienced the oppression of capitalist exploitation. Rather, ethnic groups, indigenous peoples, women, gays, the disabled, religious minorities, were the victims in a new discourse of oppressor: colonization, the patriarchy, and "Western" culture.

This conversion process was particularly evident in the way New Zealand's new middle class professionals became increasingly hostile to both the established liberal universalism of the 1940s educational reforms and to the Department and secondary schools that upheld them.[57] Now, they saw liberal Western education systems as essentially antidemocratic, combining the pervasive social control of the corporate welfare state with the seductive liberal rhetoric of permissiveness. Accordingly, the only solution was to end the power of the school in favor of a new democratic localism that would herald radical educational reform.[58]

This intellectual shift was to impact particularly on the balance between critics and defenders of the postwar secondary school. As we have seen, support for the general direction of the 1940s reforms was relatively strong within the rapidly expanding university education departments and teachers colleges during the early postwar era. Commenting on a broadly analogous situation in Australia, Alan Barcan has observed that the sociology of education made little progress until the late 1960s, with educational commentary remaining largely descriptive in nature: "...a contradictory mishmash of egalitarian ideas and progressive education."[59] From about 1967, however, Barcan argues that a combination of student unrest, the growing power of ethnic, feminist, and neo-Marxist special interest groups and the increasing polarization of educational debate, coupled with a deepening economic crisis, stimulated rapid change. One response to this nexus of social and educational currents was the adoption of the new sociology of education.[60]

Given the severity of the economic and cultural crisis in New Zealand, the growth of education and sociology as academic disciplines in their own right precipitated a decisive break with the participant-scholar tradition which had characterized the research work of earlier educational commentators such as Murdoch and Beeby. Beginning in the late 1960s New Zealand academics became progressively exposed to the new sociology at a time when they were becoming acutely aware of the education system's continuing failure, particularly at secondary level, to effectively address continuing inequalities in educational outcomes.

Cora Vellekoop's early research on the role of schools in reinforcing the impact of social class on occupational choice,[61] together with studies which focused on the structural causes of underachievement among Māori and female students, were amongst the first to expose structural inequalities in secondary education. Along with a heightened awareness of steadily worsening economic and social conditions, these widely cited studies facilitated the spread of radical ideologies among academics. Although Bates could still complain as late as mid-1978 that the new sociology of education had made no headway in New Zealand,[62] less than 15 months later Roy Nash could see indications that radical criticism was beginning to make an impact on formerly unquestioned assumptions of liberal thinking, predicting that while the New Zealand Education Department currently enjoyed a far greater degree of central control than its British or American counterparts, it too would face a growing crisis of legitimacy as the myths that surrounded the rhetoric of equal opportunity grew increasingly transparent.[63]

Once exposed to the new ideology, New Zealand academics utilized the new sociology with the passion of the newly converted.[64] By the early 1980s, D.J. Freeman-Moir was able to contemptuously dismiss the bourgeois educational "radicals" of the 1960s and 1970s as "romantic, escapists and reformist... liberals whose unsophisticated analyses of the social structure did not permit them to see that... the failure of their education system to guarantee employment put the myth under strain in New Zealand."[65] The Introduction to a widely used 1985 text, *Political Issues in New Zealand Education*[66] claimed that New Zealand was facing an educational crisis no less urgent than its economic crisis. This was because, "in the interests of Corporate Capitalism the state had determined the entire education structure," but now "the old consensus has collapsed, and a new one must be created on the basis of a correct theoretical model."[67]

In amplifying this theme, many contributors directly anticipated both the direction and the rhetoric soon to be employed by the Picot Taskforce, highlighting the extent of provider capture and centralized bureaucratic control that served vested interests and disadvantaged so many. Liberal tinkering with the existing system was emphatically rejected in favor of radical educational reform. Richard Harker asserted that if the education system was to become multicultural or even bicultural, there would have to be "a fundamental reappraisal of the structural features of our schools."[68] Introducing an early version of consumer choice rhetoric, Ranginui Walker looked forward to Māori parents, being able to "shop around" for schools which provided bilingual programs at a time when roles were declining.[69] Peter Ramsey, soon to be a Picot Taskforce member, saw schools as sites of a pervasive hegemony designed to domesticate teachers through a series of formal and informal components including regulations, a prescribed curriculum, the disciplinary procedures of teachers organizations, and the mechanisms through which beginning teachers were selected and socialized, all of which served to disadvantage many groups in society.[70]

By the mid-1980s radical sociology had come to monopolize educational commentary[71] Moreover, the rapid, late, and relatively uncritical adoption of radical revisionism, which depicted the individual as relatively powerless, tended to encourage both fatalism and inactivity.[72] Like their American counterparts a decade earlier, New Zealand radical educationalists at a crucial juncture for policymaking often adopted an anti-activist position which assumed that any disestablishment of the system would automatically lead to everything that they as educational radicals desired.[73]

Feminist pressure on the Department of Education illustrates many of the above tendencies. From the early 1970s, feminist critiques ere to impact upon the development of advocacy groups such as the National Organisation for Women and Auckland Women's Liberation. This latter organization was to provide the main impetus for the first United Women's Convention (1973), and in addition was to be responsible for the publication of the widely circulated feminist magazine, *Broadsheet*.[74] Early feminist critiques tended to focus on the practical possibilities of reform through the removal of such practices as sex-role stereotyping. From the late 1970s on, all this changed radically as feminists increasingly linked the education of girls to capitalist ideologies and structures. Family reproduction strategies were now seen to conform to "the structural requirements of both capital and the state for certain sorts of labour."[75] Studies of female teachers drew upon the notion of a "reserve army of labour" to explain why women continued to gain insufficient access to power and decision making in the education system.[76] The 1940s secondary education reforms and its central tenet "equality of opportunity" were now viewed as to a myth flawed from the outset.[77] In this context the educational bureaucracy and many secondary teachers could no longer be regarded as potential allies, but as recalcitrant pillars of vested educational interests. As we shall see in chapter five, these views were to find expression in a series of reformist reports across education, health, defense, and policing in the following decade.

It is hardly surprising, therefore, that accusations of provider capture began to be felt in policymaking circles as early as International Women's Year, 1975, when feminist pressure on the Department of Education first became evident. A Parliamentary Select Committee issued a report on the role of women in New Zealand society, and the Committee on Women, chaired by Miriam Dell, undertook the promotion and coordination of subsequent activities. In early 1975, informal discussions between Dell and Renwick led to the Department of Education and the Committee on Women agreeing to jointly sponsor the conference; Education and the Equality of the Sexes in order to promote the message that; "...both formally and informally, the education system at large (could) be either a major obstacle to change or a potential agent of change."[78]

Sector working group recommendations called for an unequivocal commitment to change. The first recommendation of the Vocational Guidance and Education for Work Working Party called for the director-general to write a personally signed letter to every teacher in New Zealand reporting on the conference. It demanded an unequivocal commitment to a "positive programme of revision," including a review of the language, curriculum, materials and methods used by teachers to ensure the elimination of stereotypical attitudes about sex roles.[79] Two major conference priorities, the establishment of a permanent monitoring committee and the appointment of a departmental officer to oversee progress reflected the emerging conviction that teachers and educational administrators were part of the problem rather than the solution. In this view, the conference was to anticipate a series of major reports and inquiries into the state sector, in addition to the recommendations of the Picot Taskforce and *Tomorrow's Schools*.

Renwick, however, seems to have believed that the two major conference priorities, the establishment of a permanent monitoring committee and the appointment of a departmental officer to service could be accommodated within current bureaucratic structures and processes.[80] He also retained control of the pace of change through his co-convenorship of the Committee on Women and Education.[81] This strategy of relatively cautious reform led to further feminist skepticism regarding the true extent of the bureaucratic commitment to radical change. Their success in influencing policy even to a limited extent, however, was ultimately dependent upon a highly centralized and directive system that alone could provide a nationwide surveillance mechanism with clear lines of accountability. In 1981, for instance, the Report *Women in Colleges*, rejected voluntary participation, making it mandatory for all teacher training institutions to report back on the measures they had adopted to reduce sexism in courses offered to beginning teachers.[82] This dilemma of dependence on a state bureaucracy they distrusted to both monitor and enforce the outcomes they deemed to be necessary was to be one that feminist and other Left-liberal groups were to share with conservatives and neoliberals. Far from easing this dilemma, devolution was to intensify it.

The Rise of the Moral Right

The growth of Left-liberal advocacy in the 1970s was paralleled by the growth of a right-wing traditionalism equally hostile to existing educational arrangements. Writing at the end of a decade of educational conflict John Codd concluded that the school curriculum had become a major political issue with expressions of concern over significant revisions in basic subjects. The result was a conflict of interest between two ideological protagonists: the "progressives" calling for an emphasis on various social and human relationships, and the "traditionalists" demanding a return to the three Rs supported by the 3Ds—discipline, diligence, and decency.[83]

The growth of traditionalist concern can be traced back to several factors. One of these was the sharp rise in youth unemployment resulting from the nation's growing economic crisis. From 1970 on the OECD was regularly posting gloomy forecasts for New Zealand's economic future, particularly as far as youth employment was concerned. By 1977 the country had the second-worst ratio of unemployed youth in the OECD, after Italy, with 52.6 percent of its registered unemployed being under twenty-one. The response of the Muldoon government was to approve, pre-employment and special training courses for young people out of work. The Minister of Education, L.W. Gandar, under considerable pressure form the Labour Opposition, called for a report on post-primary rolls as young people, many unable to find jobs, returned to school.[84] In 1977 the Department of Labour commenced a new type of program for unemployed youth—the Young Persons Training Programme (YPTP), intended largely for seventeen–eighteen-year-olds. A short time later the YPTP was joined by the School-Leavers Training and Employment Preparation Scheme (STEPS), for fifteen–sixteen-year-olds, and both programs were to be subsequently expanded as the situation worsened.[85]

In turn this crisis in youth employment caused further questions to be raised about academic standards in secondary schools. "Back-to-basics" became a public catch-cry. Behind this, lay significant shifts in social class relationships. Paul Spoonley has seen the 1970s and 1980s as a watershed for New Zealand politics similar to the 1890s and 1930s in that, during a period of major economic change and hardship, loyalties between political party and constituencies broke down.[86] Spoonley contends that the waning economic and political fortunes of the old petty-bourgeoisie—a grouping which saw its moral values successfully challenged by a new bourgeoisie which derived its cultural capital directly from the education system, and its economic security jeopardized by the rise of corporate capitalism—largely accounts for the growth of what he termed, "the radical Right" during this period. The petty bourgeoisie that formed the rank and file of this movement embraced a contradictory agenda that blamed monopoly capital, the interventionist state and what they saw as social democratic collectivism for their problems.[87] Explanations which only emphasize social class, however, may be too narrow. In 1986 Allanah Ryan characterized "the moral [R]ight" as the manifestation of populist moralism, identifying at least twenty-two organizations together with several Fundamentalist churches that were distinguished by concern for the defense of the traditional family.[88]

One of these organizations was the Concerned Parents Association (CPA). From the mid-1970s, CPA attacked what it saw as an undue emphasis on social education and social engineering in schools at the expense of the traditional curriculum.[89] Although CPA's Christian Fundamentalism probably had a relatively narrow appeal, the same could not be said of its consistent calls for more parental involvement in the day-to-day operation of schools. Moreover, CPA's increasing frustration with an educational bureaucracy which in its view supported Left-liberal education initiatives, such as the revision of school textbooks to eliminate sex bias, while apparently remaining unsympathetic to "alternative" viewpoints undoubtedly struck a responsive chord in the wider community, regardless of political and moral affiliation.[90]

Successive failures sharpened residual bitterness. Although the moral Right was able to influence some primary school committees and secondary school boards of governors to indefinitely postpone the implementation of sex education programs, they had little tangible success either in stopping the measures they opposed, or in getting the kinds of programs they wanted implemented. Yet despite the continual failure to influence policy-makers,—the unsuccessful Petition against the Homosexual Law Reform Bill, the failed campaign against the Working Women's Charter, and the abortive attempt to block the establishment of a Ministry of Women's Affairs—the moral Right remained a significant political force in the first half of the 1980s that politicians on both sides of the House were able to invoke as symptomatic of continuing public exclusion from educational decision-making processes.[91] From 1978, National's new Minister of Education, M.V. Wellington, pressed strongly for a return to formal discipline and "old-fashioned" school values. This change of emphasis culminated in the 1984 Core Curriculum Review, which stressed civic responsibilities, traditional values and an increased emphasis on core subjects within the post-primary school. As Gordon notes, this was the first public document actively supported by a sitting government to promote conservative rather than liberal curriculum views since the early 1930s.[92]

The Review, however, while winning consent from many small businessmen, representatives of traditional rural capital and Fundamentalist Christian groups, came under immediate attack from both the teachers' unions as well as from liberal educators, left-wing academics, and radical advocacy groups. In response to this offensive, Wellington claimed to be defender of parental and public interests. His bestselling book, written shortly after National's defeat in the 1984 election, was evocatively entitled, *Education in Crisis*. New Zealand education, Wellington contended, was "dancing at the edge of a volcano," "because the legitimate wishes of many parents [were] in danger of being thwarted by social engineers and theorists who no longer [saw] value in the traditional educational values of sound scholarship, sporting endeavour and cultural sensitivity."[93] Moreover, Wellington's contention that secondary school classrooms in particular were being "used to push points of view which [had] no place in a school system paid for by the taxpayer,"[94] together with his frequent assertions that educational liberals were increasingly out of touch with the views of middle New Zealand attracted a wider audience than just conservative parents. Certainly the Minister claimed that while in office, he had received "a steady flow of correspondence" in support of his stance. A graphic depiction of this view can be seen in the *Truth* cartoon of July 18, 1978 depicting "fortress education" under siege from outraged parents, produced as part of that newspaper's campaign to publicize the unresponsiveness of the education system in general and secondary schools in particular.

As we shall see, the educational conflicts and increasing polarization during the late 1970s and early 1980s was to greatly strengthen the impetus for and the legitimacy of, radical educational reform. The Department of Education and secondary schools were increasingly trapped between a resurgent Left and a reorganized Right. Unable to satisfy either they became in the eyes of both broad groupings, a personification of the problems that beset the nation. Significantly, a number of the faces of the besiegers in the *Truth* cartoon were Māori. As the next chapter will illustrate, continuing Māori educational underachievement and a growing Māori assertiveness, increasingly backed by the rise of biculturalism to a position of dominance within the caring professions stimulated radical demands for educational autonomy that were to be a key factor in the reformist agenda of the Picot Taskforce.

Chapter 4

Only Major and System-wide Reforms Will Suffice

A major difference between the educational reforms of the late twentieth century in New Zealand and educational reforms elsewhere during a similar period was the advance to political center-stage of the view that New Zealand was defined by two distinct, oppositional cultures—indigenous Māori, who were and remained the victims of colonialist oppression, and Pakeha, the descendents of British settlers. Fostered by both Māori and Pakeha activism, biculturalism was to play a major role in the general loss of faith in established educational structures that marked the 1970s and 1980s. In addition, biculturalism had a significant impact on a series of key reports, inquiries, and debates centering on the alleged failure of the public sector, particularly state secondary education, to address serious social and cultural issues. As a result, many of the key concepts of biculturalism subsequently became embedded in an emerging new policy discourse that in turn shaped the educational reforms.

A key figure in the introduction of the bicultural concept into New Zealand was Professor Ralph Piddington. As the Foundation Professor of Social Anthropology at the University of Auckland (1950–1971), Piddington established the first Māori Studies Department in New Zealand in 1952, subsequently training a whole generation of Pakeha social anthropologists and recruiting many of the senior Māori scholars who went on to lead Māori Studies Departments throughout the country.[1] Networks of crucial importance in spreading Piddington's message were established, thus spreading biculturalism, New Zealand's version of culturalism (see chapter 3), throughout the education system. Piddington and his successors subscribed to an essentialist, a historical and romanticized conception of culture as a means to social reform. As scholars they felt themselves

> constantly forced to take account of the principle of cultural relativity, and to recognize the harmonious social integration and the healthy psychological adaptation of the individual found in most of the communities which they study. They sometimes feel constrained to point out the implications of all this for the sick society of our modern world.[2]

From its outset then, both Māori activism and Pakeha conversion to biculturalism assumed a quasi-religious quality that impelled them to embark upon a process of cultural redefinition as an aspect of new nationhood identity. Māori activism, rooted in rapid postwar social and demographic changes, came to exert an increasingly effective pressure on education. By the late 1950s young Māori urban migrants had to complete three major tasks: the adoption of survival skills in an urban cash-economy; the transplantation of their culture into an urban environment; and the development of political structures and strategies for dealing with metropolitan society.[3] Postwar urbanization decreased physical separation between the races, accelerated social change, and increased awareness of disparities in housing, employment, income, and educational attainment.[4] This last disparity was particularly crucial. Although there had been a growing percentage of Māori pupils attending state post-primary schools over the decade, few Māori went on to gain School Certificate and fewer still entered the sixth form. In turn, lack of educational qualifications restricted Māori school leavers to fields of employment such as freezing works, road maintenance, factories, building trades, farming, fishing, forestry, mining, transport, and laboring.[5]

The fact that Māori activism was able to exert such a powerful influence on a whole generation of Pakeha educators lay in the fact that in New Zealand, perhaps even more than in other Western countries, the global upsurge of a global counterculture epitomized by the 1968 Paris student revolt, the Vietnam War, the anti-nuclear and anti-apartheid movements created a religious and cultural crisis in which once-dominant educational assumptions of progress were challenged.[6] As already noted, many liberal Pakeha educators came to reject Enlightenment universalism and traditional Western values. Ironically, they were now to be captivated by the symbolism and ritual of traditional Māori culture. This process was often accompanied by an acceptance of colonial guilt through a process analogous to the Christian confessional.[7]

Forging Biculturalism

Early pressure for formal recognition of Māori culture in schools was to come from the National Advisory Committee on Māori Education (NACME). At its inaugural meeting in 1955, NACME had recommended that Māori education be a special concern in teachers college preservice courses. Its political effectiveness sharply increased after April 1969, when, in response to the rapid growth of the urban Māori population and the transfer of the remaining Māori schools to Education Board control, NACME gained at least 50 percent Māori representation. Additionally, it appointed representatives from three organizations most directly concerned with Māori education; the New Zealand Māori Council, the Māori Education Foundation and the Secondary School Boards Association.[8]

NACME's 1970 report, "Priorities for Development," advocated the creation of a bicultural society. The report created widespread interest in the press, the teacher's organizations, the universities and the teachers colleges, effectively providing a blueprint for the Department of Education over the next few years.[9] NACME's 1971 Report reiterated that cultural differences had to be understood and accepted in schools. It recommended that the curriculum find a place for Māoritanga including the Māori language; and that in order to achieve equality of opportunity, special measures had to be taken.[10]

Meanwhile, the number of Pakeha educators not only prepared to embrace Māori sensibilities, but to act on their conviction that Māori communalism and spirituality could contribute to the nation's common culture by providing a counterweight to the selfish individualism and rampant materialism of Pakeha society, was increasing.[11] As early as September 1961, the New Zealand Council for Educational Research (NZCER) advocated the preparation of a handbook on Māori culture for teachers. Although a departmental memo of June 1966 suggested that this handbook simply make reference to Māori values and customs, its writer, Myrtle Simpson included a critical discussion of Pakeha agencies, from a Māori point of view. Influenced by her reading of Metge's book, *The New Māori Migration*, her awareness of Ranginui Walker's work at Auckland Teachers College, and her own recent experience in schools, Simpson radically reevaluated her own attitudes.

> I was most interested in a contribution made by a very shy Māori mother at Ohakune. She told how her small boy had come home one day after a history lesson and said, "Mother, did the Māoris eat the Pakehas?" She said he

was very upset and kept repeating "Then I don't want to be a Māori." And then I remembered a comment made by Rangi Walker of the Auckland Teachers; College staff that teachers often offended through lack of sensitivity or even lack of knowledge. He mentioned that, in dealing with the Māori wars, teachers might not always state the issues fairly and that they might put undue stress on the cannibalistic practices of some Māori warriors...here was definite support for his contention that teachers needed some advice even on matters such as this.[12]

In this revealing passage, Simpson epitomized the early role conscientization was to play in the transformation of Pakeha educators, often leading them to actively seek the conversion of others. Garfield Johnson's work at Otara College (later, Hillary College), from 1966, his mentoring by Māori such as John Rangihau and Hiwi Tauroa, and his later chairmanship of the controversial Johnson Committee, is an outstanding example of this process.[13] Another factor in conscientization was the dissemination of key academic texts by Pakeha social anthropologists such as Metge's *The Māoris of New Zealand* (1967), and *Child Rearing Patterns in a New Zealand* (1970), the latter one of numerous books written jointly or severally by James and Jane Ritchie.

Teachers unions soon became actively involved. Challenging many traditional secondary school practices, the PPTA Curriculum Review Group argued that teachers had to respect students as individuals as well as understand their interests, knowledge, attitudes, and values.[14] To this end teachers should be prepared to "thoroughly reappraise" the detrimental impact of the individualistic and competitive values that characterized society in general. It was maintained that the "interchange of ideas between school and community should be a steady dialogue" and there was a recognition that the system had largely failed Māori and Pasifika parents.[15]

Academic support for biculturalism was readily forthcoming. In 1968, Eric Schwimmer became one of the first Pakeha academic commentators to reject the concept of integration, popularizing the term "biculturalism" in his widely circulated text, *The Maori people in the Nineteen Sixties*.[16] Drawing upon Piddington's work (Piddington having contributed a key essay to the text), Schwimmer defined biculturalism as being similar to pluralism, but going beyond mere tolerance and mutual encouragement, becoming a reality only when both Pakeha and Māori cultures possessed characteristic superstructures of institutions, values and symbol.[17] Accordingly, he argued that New Zealand would only become truly "bicultural" when influential Pakeha, including educators, possessed a degree of familiarity with Māori culture that accepted the existence of two *conflicting* and equally "correct" value systems.[18] This latter claim clearly demonstrates the beginnings of the shift from the limited confines of the academy, to the wider, more politicized environment of policy and teaching.

One symptom of biculturalism's rapid acceptance by educators was the creation of a special NZCER Research Unit on Māori education. Another was the increasingly widespread involvement of departmental officers in training and in-service courses on Māori needs for teachers, guidance counselor, and vocational guidance officers. In his report, the Officer for Māori and Island Education, A.F. Smith, warned that there was likely to be little success for Māori children "unless the differing interests, values and backgrounds of Māori children (were) provided for in the programmes." Noting that NACME had emphasized the role of the teacher in dealing with Māori children, Smith called particular attention to the need for teachers to have "an understanding of how a child of another culture perceives and acts in the classroom; to know how to use the strengths of Māori children; how to draw on their experience and use teaching method (s) most suited to their needs."[19]

The NACME report also received an enthusiastic response from a key Lopdell House conference in September 1971, attended by both teachers college staff and departmental officers. The conference recommended that in the interests of social progress Māori Studies should form part of initial training for all teachers, because "it was of the utmost importance to appreciate that we have in this country a culture that has both European and Polynesian origins." Accordingly in schools, "there should be an appreciation of cultural differences and of the fact that there are alternative ways of looking at human relationships and at life."[20]

These "alternative ways" could compensate for what was seen as a fundamentally flawed Pakeha society. In 1971 Ranginui Walker claimed that, "urban neurosis resulting from a breakdown of family life and the loss of a sense of community constitute[d] one of the gravest problems of modern times." Hence, "the Māori sense of community of involvement and sharing with others could very well be emulated by the Pakeha to counter the individualism and anonymity of city life."[21] Although condemnation of European materialism on the part of New Zealand educators was hardly a new phenomenon, contrasting two distinct cultures in this way was to appeal to those desirous of promoting what was in effect, a reversed theory of cultural deficit.

The reelection of Labour in 1972 further intensified the political pressure on the Department of Education concerning the role Māoritanga might play in radically changing attitudes.[22] Labour's Minister of Education, Phil Amos, influenced the appointment of new senior officials within the department including W.L. Renwick as Director of Education, and Peter Boag, previously PPTA general secretary as director of secondary education.[23] One of the earliest indications of change was the publication by the Department of *Parent School Communication* (1973), recommending that Pakeha school principals and inspectors experience total immersion in Māori culture through signal experiences such as marae visits.[24]

Universities too, were experiencing the new mood. In November 1973, a cooperative research venture between the University of Waikato Centre for Māori Studies and the Department of Education established Te Kohanga, a preschool for Māori children on the university campus under the directorship of Dr Jane Ritchie. Although the program was initially delivered in English rather than Māori, Ritchie admitted in her 1974 report to having begun with little clear idea of a Māori curriculum, and to being humbled by parents who sang and spoke in Māori.[25] Henceforth, the program was to drastically change to emphasize Māori culture. As her partner James Ritchie was later to record, this research was to have a direct impact on subsequent policy with the institution of kohanga reo under the auspices of the Department of Māori Affairs. The University of Waikato was destined in future years to become a major center for the dissemination of biculturalism with the establishment of the Centre for Māori Studies and Research in 1972.[26]

Like many Western intellectuals, liberal Pakeha educators during the 1970s increasingly accepted the concepts of cultural relativism and ethnic particularism.[27] The first edition of Metge's widely read book, *The Maoris of New Zealand* (1967) acknowledged the debt she owed to Piddington, her former anthropology professor, while criticizing him for failing to integrate the two ideas of "emergent development" and cultural symbiosis into a single theory. The second edition of this book was an emphatic response to "the climate of world opinion, to growing disillusion with international culture built on the idea of progress."[28] In publicly admitting her failure in the first edition to properly address the spirituality of Māoritanga, Metge acknowledged both her own inadequacies, and the indispensable assistance of Māori in mentoring her to a better understanding of Māori beliefs and values. Accordingly, she found it "impossible to separate the personal and intuitive aspects of my understanding of Māori-Pakeha relations from the more objective and theoretical."[29] Written in similar vein, Anne Salmond's popular book, *Hui. A Study of Māori Ceremonial Gatherings*, appeared in 1975, followed by a second edition in 1976 and a reprint in 1983. In her preface to *Hui*, Salmond, also an anthropologist, expressed her gratitude to Metge and to the students of Māori Studies courses at Auckland University during 1970–1971.[30]

An indication of the extent to which senior Department of Education staff had come to embrace bicultural ideals by the mid-1970s is indicated in a Department Head Office submission to NACME, in early 1976. The submission conceded that schools had been used as an important element in assimilation (the original draft submission went further and claimed that schools had been a political tool). It went on to argue that although the department had moved from a policy of assimilation toward inclusion and recognition of Māori culture, there were still unanswered questions. These were as follows:

a. The degree to which we have encouraged Pakeha people to do some movement.
b. The degree to which we expect Māori pupils to match expectations of school on school terms without requiring school to match expectations of people who are their clients.
c. The degree to which we still persist on patching up such grafting on Māori language teaching without making any fundamental changes in school itself.
d. The degree to which schools change by inclusion of Māori art and culture, and providing an administrative structure based on the "whanau" concept.[31]

The period from the late 1970s through the mid-1980s witnessed a new urgency in the adoption of bicultural perspectives by Pakeha educators. Left- and right-wing critics blamed the state education system for society's socioeconomic problems; the 1981 Springbok tour and anti-nuclear issues divided the country. The Te Kooti project, directed by Frank Davis of Palmerston North Teachers College and jointly funded by the College and the Department of Education, reflected the concerns of contemporary educators to redress historical injustices and omissions while creating a new bicultural heritage for future generations. Initiated in 1978 and completed in 1981 the Te Kooti project collected written and oral material from Māori sources relating to Te Kooti, founder of the Ringatu faith.[32] Explaining his motivation for the project, Davis drew attention to his studies of New Zealand art that had led him to focus on the vexed problem of a New Zealand identity. This call for an "alternative viewpoint" of New Zealand" history suggesting "a need for a new bi-cultural mythology and its heroes, with which we can all identify" was comparable with attempts to mythologize history in the United Kingdom much earlier in the century.[33]

For Pakeha educators, the pace and depth of exposure to biculturalism was to rapidly increase during the 1970s following the introduction of regular marae-based courses for departmental officers, school inspectors, principals, deputy principals, and senior teachers on the grounds that they largely determined the policy of the school, its organization and climate.³⁴ Surviving letters of excuses for non-attendance still on file suggest that among the few valid excuses for exemption were prior engagements which could not be altered, and ill-health. One reason for this was that marae-based courses were always envisaged as serving political as well as cultural ends. An address to the Hauiti marae-based course in April 1977 was formally opened by Jim Ross, assistant secretary, Schools and Development. The first four points in his speech were recorded as:

1. Multiculturalism is out and assimilation is out.
2. We must find a New Zealand identity.
3. Many Māoris can't cope with New Zealand society which is Pakeha based.
4. Must have positive discrimination.³⁵

The program for a marae-based course from May 29 to June 2, 1977 held at Mangamuka, Hokianga, for senior departmental administrators and inspectors of primary and secondary schools constituted a rite of passage that in addition to the formal rituals of protocol and conduct, featured a bibliography that included books by Metge (*The Māoris of New Zealand*, Salmond (*Hui*), Michael King (*Aspects of Māoritanga*)and V.B. Penfold, inspector of Māori and Island Education (*New Zealanders: One people?*), together with lectures by Garfield Johnson, Alan Smith, Hiwi Tauroa, then Principal of Tuakau College, and John Rangitau, Research Fellow, Centre for Māori Studies and Research at the University of Waikato.³⁶ Although there was opportunity for participant dissent, the intensity of group feeling and the environment itself probably served to discourage it. Thus, a letter of thanks from one participant to the organizers of a course at Te Wai Pounamu College, Dunedin, claimed without irony that "it was I think most significant that there were no dissenting voices, and considering the composition of the course this was in itself remarkable. Well done indeed."³⁷ Once again, links with religious conversion are evident.

By the early 1980s, professional Māori facilitators were regularly inducting departmental curriculum officers into a selective and highly gendered Māori protocol that included whaikorero (selected officers), reply to karanga (selected women officers), mihimihi (all officers), and poroporoaki (selected offices), together with morning and evening prayers. One facilitator noted "if the task that has been entrusted to me is to ensure that ALL CURRICULUM OFFICERS acknowledge, support and are committed to reflecting our multicultural society THEN there is no better way than using a bi-cultural base as a starting point."³⁸

The notes produced by the Curriculum Development Division to accompany the Hui Whakamatatau at Taurua Marae, Rotoiti, in November 1984 attempted to define for Pakeha, the elusive fifth element "Quintessence," as:

> An all pervading element, the overall intangible bond that holds it all together. Taha Wairua—Wairua Māori perhaps. The things that cause Māori life to remain a totality. A sense of unity. Pakeha life often seems to lack that totality.[39]

For Pakeha educators venturing overseas, Māori culture was to provide a unique identity as New Zealanders. The booklet, *Te Kete Timatatanga. A Teacher Guide to Taha Māori in Social Studies*, was prepared by members of the Auckland District Social Studies Committee with assistance from the Auckland inspectors, the Māori Studies Department at Auckland Teachers College, the department's Curriculum Development Division, and the Māori and Pacific Island Advisory Team. In response to the hypothetical question "why teach Taha Māori when the school has few Māori children," the writers argued that; "New Zealanders need to know, understand and identify with the uniqueness of their own cultures. For people travelling overseas tikanga Māori provides them with a unique identity—it contributes to our wholeness of experience as New Zealanders." It followed that any teacher's lack of involvement in Taha Māori, was, "a very strong statement in itself."[40] Tuning to definitions of culture, the booklet cited Metge's (1976) view that "what is distinctive about a culture is not its elements taken separately but the way they are related to each other—their arrangement is a unique configuration," along with Greeley's 1975 definition of an ethnic group as "a large collectivity, based on a presumed common origin, which is at least on occasion, part of a self-definition of a person, and which also acts as a bearer of cultural traits."[41]

The final and arguably the most important aspect of the conversion of Pakeha educators to biculturalism, was a commitment to collective action. This element can be seen at its most urgent in *Taha Māori and Change*, a report on the proceedings of a National Residential In-service Course held at the Lopdell Centre in July 1986. Directed by Wally Penetito from the Department of Education, the course was attended by departmental officers and teachers. It aimed to select strategies for change, consider how change could be effected, and identify who was to benefit from that change. It was asserted that:

> This report is written from the pain and anger felt by the members of this course—pain, because we see the continuing oppression of Māori people in Aotearoa and the tragic results this has; anger, because those who hold

power in the education system already know well what must be done, but have not yet made significant *public commitment* to real power-sharing, and challenging of racism.[42]

The first recommendation demanded "that the Department of Education 'management plan' give absolute priority to funding programs for equity for the next five years," and that "any programme requiring funding would have to prove the involvement of and benefit to Māori and other groups disadvantaged by the system."[43]

Further recommendations called for the establishment of a Māori-nominated group to be called Nga Kaitutei o Te Haeata (The Pioneers of the New Dawn) to determine spending priorities for next five years. This group was to receive a clear undertaking by the government and the Department of Education that its decisions would be binding for a defined proportion of the departmental budget; that the senior management of the department be extended to include the numerical representation of Māori people; "and that teacher classification and promotion procedures require proven cultural sensitivity and appropriate management practices."[44]

Thus, by the mid-1980s, much of the bicultural policy rhetoric that was under-pin the Picot Report and *Tomorrow's Schools* was already in place. Meanwhile, Māori activism was having an increasingly significant impact on the wider policy environment. In the early 1980s, provoked by reaction to an incident involving the Auckland Engineering Students' Haka party and the Te Taua Community Group, Hiwi Tauroa's well-publicized report, *Race Against Time* (1982) called urgently for the elimination of cultural bias in the education system through a reexamination of educational philosophy and classroom strategies not too dissimilar to that now demanded by the school effectiveness program, Te Kōtahitanga.[45] It recommended that future training programs include as a priority, an acknowledgement of the dangers of racial stereotyping, and new bicultural education programs based on an appreciation of Māori principles and culture.[46] Schools were urged to develop a positive philosophy toward pupils of different cultural backgrounds and to respect their values. They were to encourage Māori culture in school subjects and school practices, while school-based ceremonies were to reflect Māori customs, and encourage more Māori community involvement.[47]

Even though many secondary schools appear to have attempted to implement at least some of these recommendations, they faced the familiar dilemma of meeting at least some radical demands, while allaying conservative fears. The fate of Taha Māori in secondary schools provides an illustrative case in point. Walker notes that, despite some official departmental sanctioning of the innovation, attempts by Māori to have Taha Māori included in the curriculum as a serious option were largely subverted by Pakeha teachers who attempted to delay or block its implementation, and by some schools who downgraded it by linking it with non-prestigious subjects such as technical drawing or art, thus making it difficult for children in academic streams to take the subject.[48] Furthermore, although in March 1984 Renwick assured the 1984 Māori Education Development Conference that policies were being implemented that would raise Māori educational attainment while increasing Pakeha respect for Māori culture, it was subsequently revealed that the *Review of the Core Curriculum for Schools* had in fact allocated no specific time for Taha Māori.[49] Perhaps inevitably, the initiative was, "resisted by many Pakeha people who[did] not see its relevance and who associate[d] it with 'falling standards,' and by Māori people who [saw] it as draining the already scarce resources in the Māori field."[50]

The inability of schools to meet activist demands provoked further radical proposals aimed at establishing distinctive Māori schools, of which Te Kohanga Reo was to be of particularly crucial importance. The Te Kohanga Reo ideal was first advanced at the 1980 Hui whakatauira (Māori Leaders' Conference), partly in response to Pakeha educator Richard Benton's research on Māori language loss.[51] Te Kohanga Reo was strongly rooted in cultural essentialism. It aimed to create an environment for young children which was "Māori in action and Language," while addressing the need for 'a Māori based program to stop the decline of Māori speaking people in New Zealand' without which "there [could] be no Māori culture; no unique Māori identity" It also drew upon contemporary educational theory that claimed. "it is universal truth that the first few years of a child's life is the most crucial time... for the setting of language of Social and Culture Values; of developing a concept of self; and learning to trust in oneself and others."[52]

By early 1983 there were 83 centers in operation, with 118 further centers proposed for 1983/4.[53] In requesting further information in August 1983, the Department of Education conceded that the Kohanga Reo movement was growing rapidly, and that the government had allocated substantial financial support for its development, but noted that the schools were not under Department of Education control. Hence, information could only be solicited after consultation with those Māori running it after a full explanation of the reasons why the information was required.[54] By September 1983 there were 146 Te Kohanga Reo Centres catering for 2,300 preschoolers, with 232 supervisors and 254 Māori Elders, along with 1,223 voluntary full- and part-time workers.[55] The Kura Kaupapa Māori schools were designed to build on the success of Te Kohanga Reo. Smith and Smith, analyzing the future directions of Māori educational initiatives, have argued that, as Kura Kaupapa Māori developed outside the existing state system, it implied a manifest criticism of state structures as well as constituting "a conscious and manifest resistance to Pakeha dominated educational structures and schooling processes."[56]

Despite these successes, however, many Māori and Pakeha activists continued to attack the inadequacies of mainstream secondary schooling. The Auckland Education Board noted an article in the major daily, *The Auckland Evening Star*, which commented on a report on the many difficulties of Māori pupils as encountered by teacher Maiki Marks. Marks argued that Māori language teachers believed that schools were often not working to save Māoritanga, but rather to preserve it as an irrelevancy, like Greek and Latin. Accordingly, Marks wanted schools with predominantly Polynesian rolls to have, "management committees, drawn from the community the school serves," able to, "formulate policies and select the teachers they want."[57]

These initial steps toward devolution, far from being a solution, were to become an increasing focus for further activism. By late 1984, Marshall had set up the Committee of Inquiry into Curriculum, Assessment, and Qualifications in Forms 1–7. By December, this had attracted 143 responses including submissions from the Department of Education, influential bodies such as the Employers Federation and PPTA, secondary schools, private organizations, and members of public. Of particular interest here is the submission from the Citizens Association for Racial Equality (CARE). This argued that the education system was too centralized in all its crucial decision-making procedures, and that middle class Pakeha had captured the educational process. Consequently, argued CARE, "only major and system-wide reforms will suffice to remove those inequalities."[58]

In what might have once again been a blueprint for both the 1987 Treasury briefing papers and the Picot Report, CARE proposed a comprehensive reshaping of the education system. Curriculum control was to be decentralized as a matter of particular urgency for schools where most students were Māori or Polynesian. In place of uniform prescriptions and syllabuses, schools needed broad goals and educational principles within which to work, but inside this broad structure, they were; "to be free to restructure themselves and their relations with their communities in order to become bi- or multi cultural."[59] CARE also advocated a radical decentralization of school assessment, including a shift from a norm referenced system to one where there would be a continual monitoring of student progress, together with school preparation of profile statements reflecting individual abilities and potential. By the early 1980s, therefore, Māori advocacy groups were already embracing both the rhetoric and philosophy that was to characterize the wider educational reform process.

The increasing urgency of Māori demands for educational autonomy and their major role in shaping notions of consumer choice in the interests of those considered to be disadvantaged under the existing system was also be seem in several other devolutionary initiatives at this time. Concern over gangs in the late 1970s led first to the 1979 Parliamentary Select Committee on Violent Offending, and then to the 1981 Inquiry into Gangs (the Comber Report). These were referred to Cabinet who in turn invited the associate minister of finance to coordinate ministerial suggestions on the government's response. The Prime Minister's Department and no fewer than nine state departments, including Treasury, were invited to assist. The minister of finance was later to claim that the approach taken by Treasury in formulating subsequent proposals for the Labour government that succeeded National in 1984 were based on two assumptions formulated at around this time. The first, which emerged from a reading of the reports, was that the government provision of conventional social services encountered considerable difficulty in responding to needs of young people in the "at risk" category, especially in committees that were suspicious and resisted this type of assistance. Treasury's second assumption was that the traditional way of responding to social problems—putting more money into the bureaucracy—could not continue because of the intolerable pressure it placed on government spending. According to the Minister, this assumption stemmed from the recognition that new community needs, driven by such factors as rising unemployment, could not simply be ignored.[60]

The Community Education Initiative Scheme (CEIS) was one of two pilot projects initiated in response to the report of the Committee on Gangs. Involving the predominantly Māori and Polynesian communities of Otara, Mangere, and Porirua, it was designed to assist "at-risk" youth by providing funds to be used locally for education and recreation.[61] At its outset, the scheme had to confront the tension between those who felt that local communities should use government money in the manner officialdom prescribed in order to achieve government-established goals, and those who argued that it was the community's right to establish its own goals. A compromise saw the establishment of groups such as the Otara Resource Network in June 1982 as an incorporated society with financial responsibility, but with senior public servants continuing to have a supervisory role.[62] Peter Brice, a senior educational administrator, was to be a common link between CEIS and the later Picot Taskforce—hence the scheme provided both a future administrative and a philosophical bridge between the traditional center-periphery model of educational provision, and the "consumer choice" strategies of the Picot Taskforce.

Following the snap election of July 1984, the incoming Labour government inherited CEIS. The minutes to a CEIS Interdepartmental Committee meeting in September was attended by representatives from a number of government departments including Education, Treasury, Internal Affairs, Social Welfare, and Labour. The prime minister himself expressed an interest in attending future meetings. Although the Scheme was praised for possessing the potential to manage resources more effectively, it was also felt that there were greater pay-offs for the community than just the provision of money.[63] Especially noteworthy in view of the educational financing arrangements later favored by the Picot Committee was the conviction that, given continuing financial constraints, the government had now to look for alternative methods of addressing social issues. The CEIS example, it was contended, did not require expensive resourcing. Moreover

> this was a different kind of policy. People are enabled to help themselves. The idea should be promoted with Ministers that this is an aspect of how we can deal with the whole social delivery policy. This evolves into opportunities for people to help themselves. A new kind of social thinking.[64]

The attraction of this "new kind of social thinking" is exemplified in the October 1984 CEIS meeting with key ministers in the newly elected Lange government. Chaired by Marshall, Labour's minister of education, the meeting was attended by top Labour MPs, including the Hon G. Palmer, Minister of Justice; the Hon. A Hercus, minister of social welfare and police; the Hon P. Tapsell, minister of internal affairs; the Hon. S. Rodger, minister of labour; and the Hon C. Moyle, minister of agriculture in addition to the full CEIS Interdepartmental Committee. There was considerable interest in the new approach offered by CEIS, and P. McKinley of the Treasury aptly summed up the mood of the meeting.

> CEIS was an innovative, alternative model of delivering social services. Treasury took the view that departments involved in social service delivery should seriously consider this model and consider also the ways in which their internal organisation needed to be adapted if this is to be incorporated.[65]

Meanwhile, pressure was growing for radical reform of the education system and its delivery to Māori. In early 1984 two major huis convened to discuss Māori educational needs. The first, sponsored by the New Zealand Māori Council, was held at the Turangawaewae Marae in March. Invited delegates, including members of PPTA, heard the Māori perspective on education at Huntley's Waahi Marae. "Secondary school," they were bluntly informed, was "a bitter and useless experience for young Māoris." Radical reforms were needed to help retrieve the "social disaster" of high failure rates. The entire educational assessment system was based on a white male power structure. Māori parents therefore had "no choice but to send their children into the New Zealand educational system to fail."[66]

Shortly after the huis, CARE wrote urgently to the Governor-General of New Zealand, Sir David Beattie, calling for an immediate Royal Commission to examine Māori schooling. CARE wanted the Royal Commission to specifically examine the "inability of the state system to meet Māori needs," citing what the organization deemed to be the minister of education's inadequate response to a recent question in the House as to whether the state education system was adequately meeting Māori needs.[67]

One result of the mounting public campaign for more Māori autonomy in education on the part of politically influential Māori was the increasing identification of the Department of Education as the major cause of Māori educational failure and thus, the chief obstacle to radical reform. That the department had already lost any chance of even influencing let alone restraining the tide of political opinion in Wellington was clearly illustrated in May 1986, when New Zealand's last director-general of education, W.L. Renwick wrote to the chairperson of the Waitangi Tribunal expressing the concern of his department at the Tribunal's findings. What both angered and saddened Renwick was that, although both he and his departmental colleagues had initially welcomed the Tribunal's findings regarding the place of Māori culture in New Zealand state schools, and had applauded recognition of the central importance of Te Reo Māori, the department's evidence to the Tribunal had been subsequently been subjected to improper usage and unfair comment.[68] By being selective in its citation of evidence, Renwick claimed, the Tribunal's widely reported findings had led readers to believe that the department's interpretation of Māori educational attainment was very different from the conclusions of the Tribunal.[69] The Tribunal had not even asked the department to comment on the alleged historical prohibition of spoken Māori in state schools or playgrounds, even though it clearly regarded this as an important issue in Māori educational failure.[70] As a result, readers were left with the impression that the department had either been deliberately evasive or had actually suppressed vital information regarding its historic attitude toward the use of Māori in schools and playgrounds. In his letter, Renwick concluded that the Tribunal had been "less than even-handed" in the way it had dealt with criticisms of past departmental policies, permitting the department no effective avenue of response.[71] Such reasoning, however, as we shall shortly see, had already fallen on deaf ears. The department was about to reap the reward of having become a political pariah—a position that owed much to the success of Māori advocacy groups over the preceding two decades in shaping policy discourse.

By mid-1987, therefore, as the Picot Taskforce prepared to begin its deliberations, it had available to it a wealth of evidence suggesting that Māori educational underachievement stemmed from a single underlying cause—the failure of Pakeha society and its school system to recognize Māori culture. This view was epitomized by Jack Ennis, a secondary school inspector whose *Tu Tangata* article was to be placed before the Taskforce in July 1987. According to Ennis,

> while you read these lines thousands of Māori children attending New Zealand schools are being subjected to a ten year process of schooling that very effectively and efficiently atrophies their potential growth as people. It degrades their culture and denies them the life fulfilment and expectations that most concerned Pakeha parents expect and demand for their children.[72]

Ennis slated the Department of Education's record with Māori students as "a dismal failure," citing former departmental officer, D.K. Royal's claim to a gathering at the Orongmai Marae in December 1985 that both the system and structure of education were wrong.[73] Although he held the system largely to blame, Ennis also argued that his own experiences in Porirua secondary schools suggested that teachers regarded Māori educational retardation as the norm and often "put down" Māori students.[74] By the time that the Picot Taskforce met, therefore, the Department of Education and secondary schools were already being seen as the major cause of Māori educational underachievement, while radical Māori demands supported by many liberal Pakeha educators, were already calling for devolution.

Chapter 5

The Best Kind of Accountability

The first chapter in part 2 of this book examined the dual economic and cultural crisis that was to impact upon secondary education from the late 1960s on, leading to increased polarization and the rise of radical advocacy groups from a cross the political spectrum. The next chapter focused on development of biculturalism and particularly the impact on secondary education of demands for Māori educational autonomy. This chapter, the last in part 2, centers on the rapidly changing policy environment and the shaping of a common policy discourse around radical education reform, particularly as this was manifested in a number of key conferences, reports, and inquiries.

Demands for Radical Reform

Important as they were in both molding and polarizing public opinion on secondary education, advocacy groups of various kinds were far from being the only source for widespread disquiet over secondary education. During the 1970s and 1980s there were increasing calls for radical reform across the state sector that foreshadowed the Picot Report and *Tomorrow's Schools*.[1] The central versus local issue in New Zealand education had long been problematic for secondary education. From the late 1960s, however, centralization actually increased as the Department of Education bureaucracy sought to counter ongoing public criticism. The creation of the Curriculum Development Unit (CDU) within the department came about through a recommendation of the Currie Report, its first Curriculum Development Officer being appointed in 1963. Furthermore, although a highly centralized process of curriculum-making had existed since the nineteenth century, the increasingly techno-scientific bias of educational innovation from the 1960s on meant that a comparatively small group of "experts" within the inspectorate and department came to exert a strong influence.[2] Consequently, the development of new approaches to social studies (1976) and English (1977) were innovations imposed by relatively small groups of curriculum designers, operating largely outside the sphere of public consultation.[3] The latter document in particular, influenced by overseas theorists such as Basil Bernstein was viewed with considerable disquiet by conservative parents and politicians as well as attracting criticism from some secondary teachers who felt that it heralded a reorientation of the whole curriculum toward social engineering goals.[4]

Unfortunately for the department and for liberal educators, these new curriculum innovations fuelled further public unease at a time when parents were becoming increasingly aware of their collective ability to influence the indicative planning exercises that were to be increasingly important for secondary education. The National Development Council established in 1968, reported directly to Cabinet. A raft of newly created sector councils reported to NDC, including the newly created Advisory Council on Educational Planning (ACEP). Following the abolition of NDC under the Labour government of 1972–1975, ACEP reported directly to the Cabinet Committee on Policy and Priorities.[5] As governmental concern over the rising costs of educational development mounted during the early 1970s, ACEP became the steering committee for a planned Educational Priorities Conference that in due course became the Educational Development Conference (EDC).

EDC proved to be a massive undertaking, involving widespread participation across the community. Through the many reports and discussion booklets submitted by its various working parties, EDC contributed extensively to existing educational literature. One particularly significant outcome was the 1974 Nordmeyer *Report on the Reorganisation and Administration of Education*. This recommended that primary school committees be enabled to take part in staff appointments thus bringing them into line with secondary school Boards of Governors. The Nordmeyer Report also drew attention to a remit passed by the 1973 Labour Party Conference that proposed the abolition of education boards.[6] The last of the EDC reports was the *Report of the Advisory Council on Educational Planning*, which recommended that the government give more discretion to school controlling authorities regarding funding, curriculum, and staffing matters.[7]

EDC also acted as a clearing house for reports by regional committees on the outcomes of public seminars and activities. An estimated 60,000 people took part, and some 8,000 submissions were received, with press and radio contributing to public discussion. Commenting on the overall impact of EDC, Frank Holmes claimed that marae-based Māori gatherings and secondary student seminars were among the most successful and influential meetings held.[8] Given the upsurge of Māori educational criticism focused particularly on state secondary schools, this development was to be highly significant for the future. Thus EDC's greatest achievement was "to educate a very large number of both the laity and professional educators about current educational problems, about prevailing attitudes toward education held by different groups, and about the possibilities of educational progress through greater interaction between educational institutions, parents and other groups in the community."[9]

EDC had an indicative impact on the future reform of education, revealing as it did considerable community frustration with, and alienation from, an educational bureaucracy widely perceived to be out of touch with current aspirations, especially those of Māori, women and other disadvantaged groups.[10] It illustrates that, "from the 1970s in particular, there was a steady build up of dissatisfaction with the existing system and a series of reports recommending reform in several areas which foreshadowed and were almost identical with those subsequently recommended by Picot."[11] As far as secondary schools were concerned, additional problems were revealed concerning a lack of financial discretion individual learning institutions possessed in a centrally controlled system, ongoing structural problems involving the department and secondary school boards, and the general difficulties of co-ordination between different parts of the education system, especially those between secondary schools, post-compulsory education, and the workplace.[12]

In particular, the steady erosion of the powers of secondary school boards of governors to departmental authority had become clearly evident by the 1970s.[13] The McCombs Report on secondary education, *Towards Partnership* (1976), identified serious flaws in existing arrangements and considerable public dissatisfaction, particularly at the local level. The report contended that real cooperation depended on lay people being able to participate in decision making rather than simply being told what was happening. It recommended immediate government action to improve relationships between teachers, parents and local communities.[14] Several McCombs Report recommendations were later adopted by the Picot Taskforce and by *Tomorrow's Schools*, including recommendations that the majority of boards of governors be parents; that boards needed a better system of reporting to the community; and that boards express community views about the curriculum rather than simply leaving it to the school principal.[15]

In response to the McCombs Report, the director-general of education set up a task force to study departmental staffing and functions. Chaired by the department's own Divisional Director (Special Duties), H. Egdell, the four person committee included three members who were either departmental officers or inspectors, with one seconded SSC member.[16] To a large extent, as with previous attempts to reform the system, the Department of Education was able to temporarily blunt much of its impact. Nevertheless, the tacit admission that serious structural problems in secondary education administration actually existed gradually permeated even internal critiques. In 1977 the then principal of Makora College, Noel Scott, completed a survey of New Zealand secondary schools for the Department of Education that identified a number of problems then facing secondary schools.[17] A frequently voiced complaint concerned the lack of a clearly defined national philosophy of education under which, Scott believed, individual school interpretation could shelter. Scott also observed that the EDC had highlighted a wish for more clearly defined national guidelines. Anticipating the Picot Taskforce's call for guidelines that would find local expression in school charters, he suggested that any future body charged with drawing up such guidelines could well be drawn from a wide cross-section of professional and lay people.[18] Highlighting a key issue later identified by the taskforce, Scott pointed to the historic conflicts and growing antagonisms between various sections of the Department of Education and secondary schools that had created sharp barriers between the policy-forming and policy-implementing sections of the system with the result that, "the deep, frequently voiced criticisms, even antagonisms evident in dealings between various sections of the Department of Education and schools, seem to erode the very foundations of the type of constructive, supportive, unified efforts which education deserves."[19]

Similar problems were highlighted in a Department of Education draft statement for the OECD Examining Panel sent out to New Zealand to review educational policies in early 1982.[20] The statement characterized education as a highly political activity, with educational lobbies being

> among the best organised and most persistent in the country. They are well connected with the media. Public opinion is, as a result, constantly massaged.[21]

This revealing statement conceded that secondary schools faced problems relating to changes in the family, ethical/cultural diversity, public/private responsibility issues, economic restructuring, technological changes, shifting employment patterns and public expenditure constraints. At the same time competing interests within education meant that educational progress was often slow, occurring over a broad front, hence effective consultation was increasingly difficult to achieve.[22] Moreover, it was conceded that there were contradictory arguments—some asserting that the department consulted too much with consequent wastage of money and time, and others complaining that consultation was too limited.

Accordingly, the departmental statement favored incremental modifications to existing arrangements, including the expansion of committee representation, and curriculum reforms serving the needs of non-academic as well as academic students, rather than any radical reform.[23] This compromise was hardly surprising. The introduction to the document, written by Renwick himself, noted that the departmental statement, together with visits the panel would make during their three weeks New Zealand stay, would form the basis on which the panel would write its report. This report along with the Education Department's Statement would then form the basis for the published OECD document.[24] Predictably this last document was highly complimentary of New Zealand's educational achievements, referring to a high degree of teacher professionalism and public confidence in a system that, in the view of the OECD examiners, was efficiently run.

Despite this endorsement, however, the OECD Report obliquely critiqued the longstanding tendency for the post–World War Two New Zealand education sector to priorities purely educational values over economic and instrumental goals. Noting that the Currie Commission was one amongst several post-war educational reports that had seen fit to devote little space to the relationship between education and the economic system, and citing the possible impact a greatly changed contemporary employment situation might have on thinking, the OECD Report warned prophetically that

> to suggest that closer links be forged between education and economic planning, so as to enable the contribution that schools, colleges and universities make to social and economic development to be widely discussed and made more explicit, is in New Zealand seen as representing a particular political viewpoint rather than a perspective on educational policy that deserves due consideration by all sections of the population.[25]

Once again, the Thomas Report's emphasis on the intrinsic value of a well-rounded secondary education was being brought into question. Moreover, attempts by the educational bureaucracy to contain criticism of either the way the system was being run were now becoming increasingly difficult in the rapidly changing political climate. For example, the increases in teacher militancy during the 1970s, while alienating some parents, had led the PPTA to support the power of local boards as employers of teachers as opposed to the power of department and minister.[26] In this context, teachers and parents were at least temporarily united against encroaching central authority. In particular there was a growing sense of powerlessness in the situation where secondary schools boards of governors often approved decisions made elsewhere, by the principal or by the Department. Hence they were "like a small child playing with a ball on the sideline at a rugby test match—it makes no difference whether s/he is there or not."[27]

Increasing pressure for more local autonomy inevitably had a political impact. In the run-up to the 1984 election, education became caught up in the political campaigns of both Labour and National to capture the "highground of public opinion."[28] The years between 1984 and the reelection of Labour for a second term in 1987 were to provide a crucial indication of what lay in store. Under considerable public and political pressure the minister of education, Russell Marshall, promised wider public consultation including a review of senior secondary school and assessment procedures. A select committee of inquiry into the quality of teaching had identified a cumbersome administrative system, provider-capture, and lack of teacher accountability as factors undermining teaching quality (see 155–157). Heightened political interest in education during the early 1980s was also reflected in the growing importance the Opposition National Party accorded its new education spokesperson, Ruth Richardson, the redoubtable member for Selwyn. When Richardson was appointed education spokesperson in early 1985, the position was widely regarded in her party as being a political graveyard. Richardson, however, rapidly made educational issues a major part of the Opposition's attack on the policy of the Labour government. In July 1985, the Labour and Education Committee's report on the Education Amendment Bill was presented. This bill, eventually passed as the Education Amendment Act, 1985, sought to make legal provision for parents to be consulted about the health syllabus before the school drew up its program, give managing bodies of schools power to determine whether a sex education component was to be included in the health program, and give parents the legal right to withdraw their children where such a component was taught. Although Labour claimed that these arrangements represented the middle ground between the various views laid before the Committee, Richardson was quick to contrast what she regarded as two opposing sets of submissions. The first set, represented by the New Zealand School Committees Federation, the New Zealand Parent Teachers Associations, and the New Zealand Education Boards Association emphasized the need for a voluntary partnership between family and school. The second set, epitomized by the submissions from the two teachers' unions merely paid lip service to this view, implying that the final decision about curriculum content should be taken by the teaching profession in partnership with the educational bureaucracy.[29]

The renewed vigor of National's attack had further political and policy ramifications. In July 1986 a committee set up to undertake revision of the 1964 Education Act once again emphasized the need for more flexible legislation. Two issues the committee raised were to be of considerable relevance for secondary schools. First, directly anticipating the Picot Report, the committee recommended that the Department of Education become a "Ministry," and the director-general of education, the "Secretary of Education."[30] The committee was also desirous that any new legislation permit further delegation of power to local authorities, where appropriate.[31] Moreover,

> secondary school boards should have a majority of parent's representatives and should have a student representative. Those few secondary school boards at present governed by their own nineteenth century acts are to be brought under the Education Act'.[32]

Meanwhile National's attacks were having an impact upon an increasingly nervous Labour government. By early 1987 Education Minister Russell Marshall was indeed "a Minister under siege."[33] His response was to announce to his political colleagues an intention to review the compulsory education sector. This move, supported as it was by the Department of Education, has been seen as "a pre-emptive strike" with the intent of forestalling more radical initiatives.[34] Clearly, public and political sentiment toward education was hardening. In April 1987 the popular magazine *Metro* featured a cover story by Carol du Chateau. Entitled "The Lost Generation," it portrayed students in New Zealand schools as having been captured by a liberal-progressive teaching profession led by radical feminists and gay activists.[35] In the same month Marshall, acutely aware of the Cabinet's increasing receptivity for thoroughgoing educational reform, admitted to a Waikato University graduation ceremony feeling uneasy over "present market-economy thinking" when applied to education. He claimed to feel comforted by a belief that this was an aberration, but conceded that the questions market adherents raised required a serious response.[36]

The long-awaited report of the committee to review the Curriculum for Schools, requested by Marshall back in late 1984, also served to increase pressure.[37] The report recommended a national common curriculum for all schools, from new entrants to form 5, thus encompassing all the primary years and the first three years of secondary schooling. Although this was to provide a broad general education consisting of national curriculum principles and three interrelated aspects of learning: knowledge, skills, and attitudes and values, each school was to have "responsibility to develop a school curriculum which is consistent with the national common curriculum," with responsibility falling jointly on students, teachers, administrators, parents and the community."[38] In phrases that anticipated the Picot Taskforce's conviction that community involvement in all aspects of the school could solve the problems of alienation and inequity, the report observed that parents felt isolated from the day-to-day running of the school and the school programs;[39] that those elected to manage schools did not fully represent the community;[40] and that the community clearly had a role in staff appointments.[41] Foreshadowing the charters later recommended by the Picot Report, the Committee contended "that all communities should, through their elected representatives, share in writing a school profile and a statement of community expectations."[42]

Events, however, were moving rapidly beyond Marshall's control. An sample opinion poll of 2,500 people taken on May 20 and subsequently published by Radio New Zealand and the *National Business Review* revealed that only 25 percent thought education was working well, with 17.7 percent believing that it was working "quite well." By contrast 28 percent thought it was working poorly and 14 percent, "very poorly." Responding to this survey, a *Dominion* editorial warned that there was "a strong public feeling that the education system [was] not working."[43] The editorial went on, however, to caution that current government policies that had widened social and economic gaps while creating high unemployment had fuelled these concerns with the result that education had become a scapegoat. In response politicians shifted the blame onto teachers for not turning out marketable products that met demands for flexibility in the new economy. Claiming that right-wing think tanks in New Zealand and overseas were the most prominent in advocating a shift in taxpayer dollars from the education system to the individual, who would then purchase education services from a variety of free-market options, the editorial observed that

> the National Party has picked up on this vogue thinking in its education policy. It is strong on sentiment, heavy in rhetoric and short on detail. The right-wing ideology is watered down, but it is not clear whether that is to make it more palatable to the public or whether National can't quite make up its mind. Either way, the direction is unmistakable.[44]

The direction in which the Lange government was now headed was likewise, becoming unmistakable by the week. Immediately following the general election, Lange outlined the broad approach of his government to education over the next three years to PPTA's Annual Conference.[45] In a somber address, Lange observed that because New Zealand's economy and society were changing rapidly, education needed to reflect this fact. Ominously, he warned his audience that there was no longer any assurance that a buoyant labor market would absorb young unskilled people. Hence, it was "no accident" that the minister of employment was also the associate minister of education, with responsibility for both tertiary education and continuing education, especially given that these sectors were "inextricably bound up with demands of labour market."[46]

Turning to the newly set up Picot Taskforce, Lange sought to reassure his audience that the purpose of the review was not to be a cost-cutting exercise, but rather to look for gains in efficiency. He did, however, mention the significance to the taskforce of the recent review of school zoning, the continuing issues raised by the Curriculum Review, the findings of the Scott Report, and the Royal Commission on Social Policy, which in turn necessitated his own need to keep in close touch with current thinking on education.[47] Responding to the recent Radio New Zealand and National Business opinion poll, Lange argued that it was his responsibility to find out "why public confidence has sunk to such a low level."

In tones that clearly reflected the current provider capture arguments the National opposition was employing, Lange expressed concern that the PPTA had opposed proposals for the introduction of merit pay and short-term contracts as well as critiquing the Scott Report for threatening to deprofessionalize teachers through proposing more stringent external forms of control. However, he stressed that "real accountability (was) not...some kind of burdensome necessity." Rather "it (was) at best an assurance of professional responsibility."[48] Warning that there would be no more "limitless resources." Lange referred to the need to make choices and to set responsibilities.

> That is a test of judgement as stern as any you are called on to make in the classroom. The public is entitled to hold politicians, administrators and teachers to account for that judgement and it is my intention to see that we are all properly held to account.[49]

If anyone in secondary education had mistaken this message, then they would have been under few illusions following Lange's equally emphatic address to the New Zealand Secondary School Boards Association's Biennial Conference just two days later. In this address, Lange stated that although he had not yet had an opportunity to meet the taskforce members, he hoped soon to do so when

> I shall encourage the members of the taskforce to be radical in their approach. I would like them to examine every option there is in educational administration, however unusual or untoward it may appear at first.⁵⁰

The prime minister conceded that there would be limits to radical reform. The government was committed to public education, and any changes the taskforce recommended would be undertaken within a public sector context. Hence, he summed up the taskforce's task as

> the development of an education administration in which both the public and the people who work in education can have confidence. By that I mean that education administration must be accountable for its performance.⁵¹

Lange's views were not without contradictions. On one hand he looked forward to mechanisms which would require the government to be more active in its determination of priorities and more open and accountable in its planning for education.⁵² However, because it was "unrealistic" to hold the minister or director-general to account for what happened in day today administration, responsibility, and accountability had to be devolved.⁵³ Recalling the criticisms of education recorded by OECD examiners in their 1982 report, Lange conceded that public consultation had not always reflected public views, but instead reflected what could be broadly described as the views of the educational establishment.⁵⁴ On the other hand, however, he claimed to be unconvinced that the answer to accountability lay in more devolution to community representatives'.⁵⁵ The strength of bureaucratic and professional interests might predominate if there was any conflict. The interests of the board and staff of a particular school might not be those of whole community, which would then have no effective redress.⁵⁶ Thus, "the best kind of accountability" was "being seen to do a good job." Those who failed in their responsibilities would eventually face dismissal.

Meanwhile public controversy over provider capture in education continued unabated. In highlighting the issue both in the House and in the media, Richardson had utilized the title, *A Nation at Risk*, to evoke the images of parent power coupled with the rhetoric of excellence that the United States report of that name had promised nearly five years before.⁵⁷ Here, Richardson was able to tap into a growing body of opinion that saw secondary schooling as inadequate preparation for the world of work in the context of an economy that would have to become increasingly flexible in a global marketplace. In June 1986, a report prepared by the New Zealand Vocational Training Council for the OECD in response to that organization's call for member states to reexamine the role of education and training in economic performance and development. Although the 1982 OECD Report had dealt somewhat obliquely with the links between education and the economy, the Vocational Training Council's report owed much to the 1984 British report, *Competence and Competition*. Like the British document, its New Zealand counterpart noted how Germany, the United States, and Japan had enhanced their economic effectiveness, including a more rapid response to the changing global marketplace, through a combination of vocational education and training initiatives that began in the secondary school and continued into the post-compulsory sector. It was concluded that there was evidence of an increasing body of opinion in New Zealand that would support similar reforms.[58]

National's 1987 election manifesto, also entitled *Nation at Risk*, promised bold, innovative and far-reaching educational reforms. These included a modular national certificate that would replace earlier vocational qualifications and an emphasis on the maintenance of national standards, coupled with a significant reduction in the role of the state in education to quality control and funding with an emphasis on the rights of consumers rather than providers.[59] The brief for the Picot Taskforce was fast being written.

Education and Treasury

Increasing public and parliamentary concern over both the control of education and its aims, particularly at secondary level were but the visible tip of a threatening political iceberg. Chapter two outlined how growing economic concerns brought a new generation of graduates into the New Zealand Treasury who went on to effect decisive changes in both its institutional philosophy and its approach to politicians. By 1977 Noel Lough had become Secretary to the Treasury. McKinnon has related how "the conflict of the early 1980s was not only a matter of individuals with new ideas gaining influence—it was also an episode in the more than century-old contest over who should guide economic management—Treasury experts or the politicians."[60] All this was to be of direct relevance to secondary education as the relationship between the Education sector and Treasury steadily deteriorated. Reflecting both financial issues and a growing Treasury willingness to critique contemporary social policy, the relationship between the two began to sour dramatically in the late 1970s, to reach crisis proportions by the mid-1980s.

Although the 1960s had seen a continuous growth in educational expenditure from 11.2 percent of net in 1961, to 16.2 percent by 1971, the proportion of government expenditure on education began to fall after this date. Even as it attempted to steer a middle course through a newly radicalized and increasingly polarized educational discourse, the Department of Education from the mid-1970s on was attempting to counter growing criticism from Treasury and SSC critics who sought to cut educational costs, while increasing accountability in a system that demonstrably seemed to be failing the public. A key factor in the growth of the new public-sector domiciled economic rationalism was the adoption of Canadian-inspired public sector planning strategies as a direct response to the fiscal crisis that followed the "oil shocks" of 1973 and 1974.

By early 1975 an embattled Labour government, headed by the new and relatively inexperienced W. Rowling, was confronting a major economic crisis. Facing impending election defeat at the hands of an invigorated National opposition under the leadership of ex–minister of finance, R.D. Muldoon, Labour scrambled to control expenditure. Accordingly, in January 1975, Treasury commenced development of a Computer Forecasting System (FFS), designed to service the forecasting needs of government departments and to enhance both planning and control of available resources.[61] With the reelection of National in November, the stage was set for still more drastic reforms.

The expenditure of all departments of state was still at this juncture subject to approval by the Committee of Officials on Public Expenditure (COPE) representing seven departments, including Education, and Treasury. In August 1977, however, a report from the Task Force on Economic and Social Planning, prepared by Treasury and approved by the Chairman of the Planning Council, Sir Frank Holmes, proposed that revised machinery for planning government expenditure be established. Drawing upon recent Canadian experience with Public Sector Planning, the report envisaged that the revised COPE would henceforth consist of only four control departments. Treasury, supported by the minister of works, the State Services Commission, and Trade and Industry would hold permanent seats and voting rights, with the fourth vacancy rotated every two years amongst three other designated departments; these being in the first instance, Agriculture and Fisheries, Social Welfare, and Defence. In the short term at least, Education was to be left without direct representation. In addition, the Planning Council would have the authority to take recommendations regarding priorities and allocations directly to Cabinet, thus by-passing COPE completely.[62]

The Education Department representative to a stormy meeting at Treasury reported back to Renwick; "as we suspected departments gave the proposals a rough passage and both Treasury (Mr Cliff Terry) and his colleagues and Sir Frank Holmes and his colleagues were pretty much on the defensive." The representative also reported that; "one questioner asked Sir Frank and Mr Terry who was running the country?"[63] Other objections centered on the viability of COPE especially given the increased role of Treasury and the Planning Council, the representation of departments not initially included on the committee, and the new importance placed on representatives with financial expertise rather than any association with policy formation in the various departments.[64]

While the trenchant opposition of many government departments led to the temporary shelving of Treasury and Planning Council Plans, Treasury continued its unremitting drive for further economies. A Treasury Report highly critical of auditing procedures in all government departments was submitted to Parliament in May 1978.[65] Subsequently, however, the Department of Education was singled out for particular attention. In March 1978, a Financial Management Review team from Treasury completed a review of the Education Department. The team informed Renwick that the recent Treasury report was "almost wholly applicable to your department." In the team's opinion, "the most urgent matters for attention (were) the upgrading of internal audit within the department and the service that team could then give, together with management accounting section, in raising the financial accountability and responsibility of boards and schools."[66] More damaging still from the public relations perspective, was that the team's conclusions were leaked to the media, resulting in extensive radio, T.V. and press coverage. The extent of political fall out was exemplified in hard-hitting front page headlines such as that in the *Dominion* of June 28, 1978 entitled; "Where Your Money Should Not Have Gone."[67]

In July, Renwick wrote to the minister complaining that the comments about lack of financial accountability and low auditing standards was "extremely broad and reflects harshly on those education controlling authorities which are managing the resources under their control in a competent fashion."[68] Although accepting that there was a need for improvement in financial management by education authorities generally, Renwick argued that the report was not so much a criticism of the department as it was a reflection on the ability of a depleted staff to carry out its functions.[69] In September, he told the associate minister of finance, Derek Quigley, that Treasury had reported adversely on a proposal submitted by the department to increase its grant to the Vocational Training Council, complaining that the Treasury stance had not been discussed with departmental officers.[70] In November, he informed the Secretary of the Treasury that the mammoth task of reporting on expenditure by the required deadline had not been helped by the delay in receiving expenditure statements from Treasury.[71] In March 1979, Renwick commented more caustically that an adverse Treasury Report on the establishment of rural centers for the Psychological Service was "a superficial statement based on a simplistic understanding and interpretation of the work of the Psychological Service and the facts available." It was a false economy that failed to recognize the needs of children as a first priority.[72]

Worse was to follow. As it transpired, Education was one of the last two government departments to get FFS, and it was not until 1979 that Vote Education was presented in a new format focusing on "components" instead of programs and activities. A particular feature of this more detailed breakdown of expenditure was the separation of reasons for variation in the department's forecasts into those due to price changes, and those due to real changes in the level of operation.[73] This innovation greatly increased the ability and the willingness of Treasury officials to scrutinize educational expenditure on a regular basis, often employing questionable tactics to secure information. In February 1980, for instance, the department complained to Treasury that their officers had made approaches to junior departmental officers who were not in a position to give full and accurate information, and thus Treasury reports had not always conveyed an accurate picture of departmental operating procedures.[74]

Itself under pressure from a cost-conscious government, Treasury continued to express considerable concern at the delays inherent in the COPE exercises which in its view lead to the promulgation of draft estimates requirements too late in the year.[75] In March 1980, Quigley prepared a confidential report to Cabinet on public expenditure. Citing both the Canadian Sector approach and the recent initiatives in the United Kingdom under the Thatcher government, the report aimed to achieve a greater degree of control over government expenditure than allowed for under the existing COPE, which tempted departments "to adopt an expansive viewpoint when determining their needs."[76] In contrast, Quigley argued that; "What is required is an incentive that will induce departments to examine their expenditures not in terms of their desirability but rather their necessity and priority in the light of our policy objectives and the other resources available."[77]

A fortnight later, Cabinet agreed to the setting up of the Officials Committee that included Renwick, to examine the recommendations of the Quigley Report. Renwick in particular, was highly critical. In a letter to the minister, he questioned the need for change for the sake of what he saw as largely illusionary advantages, and pointed out that any top down expenditure allocation process would pose serious problems. He also observed that, since 1976, the education vote had been subjected to extremely rigorous examination, and he questioned the ability of Treasury evaluation teams to make educational decisions.[78] Nevertheless, in July 1980, Treasury announced plans to dissolve COPE completely. In its place, three or four panels of officials would report directly to the minister of finance on the future costs of existing policies. One panel would consist entirely of Treasury officials, but each of the others was to have Treasury representation. Once again the proposals met opposition from the Department of Education. In August 1980, it was pointed out to the minister that three meetings of the new Officials Committee had not produced consensus, hence the department could not believe that the Treasury proposals would save time.[79] Despite opposition, however, Treasury had succeeded by the early 1980s in requiring all departments of state to confirm to a strict calendar as far as expenditure was concerned, with Treasury intimately involved throughout the process.

Despite Renwick's energetic defense, worse was to follow after National was returned to the Treasury Benches following the 1981 Election. By early February 1982 Treasury was warning that the economy had not performed as well as that of many other countries in terms of employment opportunities, balance of payment, equilibrium, economic growth, price stability, or distribution of income. Treasury considered the nation's internal deficit to be around 8 percent of GNP, nearly twice as high as it had been a decade earlier, and about twice the OECD average. Cabinet was advised that the deficit needed to be urgently reduced by imposing savings across the state sector. In response, the government immediately imposed 3 percent savings on all government departments, including Education, with strong hints of further cuts being raised in the national press.[80] All new policies or alterations to existing policies leading to increase in expenditure together with any changes in timing, distribution, or source of expenditure were to be subject to Treasury and Cabinet scrutiny.[81]

Continuing Treasury interest in the education sector was to have profound implications for the Education Department immediately prior to the establishment of the Picot Taskforce. From July 1984, the Lange government attempted to deregulate the economy on the grounds that this would make it more competitive and less distorted.[82] The need to comply with government economic policy placed the department in an invidious position. Various advocacy groups vigorously competed for scarce resources.[83] At the same time, education was stigmatized as a wasteful spender, leading to further criticism of the department's financial management. Most important of all, the department was to be increasingly excluded from the policy decision-making loop, but at the same time obliged to work closely with Treasury officials in a climate where education was increasingly linked to economic performance.[84]

The New Policy Rhetoric

Perhaps the clearest indication that radical educational reform was now inevitable was the growing ascendancy of a new policy discourse in governmental and bureaucratic circles. From the late 1970s there had been an increasing tendency for politicians to utilize public inquiries, thereby circumventing insider advice through a direct appeal to "public opinion." During these years the National government under Prime Minister R.D. Muldoon and its successor, the Labour government of David Lange actively practiced this highly politicized strategy. The National prime minister utilized the public inquiry mechanism to critique both individuals and institutions, circumventing the existing processes to appeal directly to both the media and the general public. In 1978, the Colin Moyle Affair led to a full commission of inquiry into an alleged breach of security involving a police file on Moyle, a senior Labour MP.[85] A year later the long-running and sensational Arthur Alan Thomas murder case was reopened, attracting intense public and media speculation centered on an alleged police cover-up, and leading eventually to two inquiries that pointed to serious flaws in the Crown's case against Thomas.[86]

The new culture of distrust soon spread. The 1980 crash of an Air New Zealand DC10 aircraft in the Antarctic with the loss of 257 lives resulted in a Royal Commission being set up under the Hon. P.T. Mahon, an Auckland High Court Judge. Mahon concluded that there had been on the part of airline official witnesses he spoke to; "a pre-determined plan of deception," that was "clearly part of an attempt to conceal a series of disastrous administrative blunders." "I had to listen to an orchestrated litany of lies."[87] This last hard-hitting phrase was subsequently to become part of the nation's political folklore. Thus, by the time the Labour government assumed office in 1984, not only were both the police and the national airline, both established icons of public trust, discredited in the public eye, but there had also been a precedent established whereby politicians appealed directly over the heads of permanent public service heads to a wider reading public.

This pattern of public inquiry accompanied by media speculation and the injection of an evocative, pungent, and appealing rhetoric into the policy environment was to be further extended under the Lange government. In 1985 for instance, the government appointed a committee of inquiry into defense chaired by the Hon. Helen Clark, a future Labour prime minister. Virtually ignoring the views of defense chiefs and foreign affairs bureaucrats that had previously helped shape New Zealand's traditional defense policies, the committee instead drew upon what it identified as a considerable depth of public concern about nuclear weapons, instead declaring support for the more radical view that conventional defense arrangements no longer had any relevance.[88]

It was developments in the health sector, however, that were to prove particularly indicative of what was in store for education. Provoked by an article in the serial publication *Metro*[89] condemning the treatment of cervical cancer at National Women's Hospital written by Sandra Coney, a feminist journalist, and Phillida Bunkle, later to be a New Labour cabinet minister, a Committee of Inquiry was appointed in June 1987 headed by Silvia Cartwright, a district court judge. The resulting report was highly critical of existing professional standards and safeguards. In hard-hitting phrases that were to directly anticipate the rhetoric adopted by the Picot Taskforce, the Cartwright Committee highlighted the inadequacies of existing system to meet current consumer needs. The committee recommended the external audit of clinical standards, the development of quality assurance programs involving clients, a patient advocacy system that was to be independent of the hospital, and the imposition of sanctions for failure on the part of any health professional to comply with these statutory obligations to patients.[90]

The influence of the Cartwright Report was soon to be felt across the entire health sector. In early 1987, shortly before the setting up of the Picot Taskforce, the Gibbs Taskforce was set up under the auspices of the ministers of health and finance. In its identification of problems, its subcontracting of research to private organizations, its rhetoric and its proposed solutions, the resulting Gibbs Report provided a further blueprint for state sector reform. Like the Picot Taskforce, the Gibbs Taskforce began its deliberations with a list of perceived problems, beginning with equity. System responsiveness was identified as a key issue, with the taskforce gaining the definite impression from submissions that, "people considered health and medical care too important to be left entirely to the decisions of doctors."[91]

Citing the Cartwright inquiry as having raised numerous questions about health professional's attitudes toward patients, the Gibbs Report claimed that medical behavior and control systems in hospitals lagged far behind community expectations. Hence, there was a need for corrective measures for unprofessional behavior as well as redress for patients. It was further claimed that there was growing community disenchantment with an apparent "inability to bring about change, to have their complaints given weight, or simply to be treated as intelligent adult people." Complaints from women's groups and Māori about lack of cultural sensitivity in system exemplified in the alleged attitude among many professionals that, "they don't know what's good for them, were particularly highlighted."[92]

In a further move that foreshadowed the approach adopted by the Picot Taskforce, the Gibbs Taskforce commissioned a private firm, Arthur Andersen and Company, to examine the efficiency of the sector, These investigations revealed a general inattention to good management, leading the taskforce to highlight this issue as the main problem.[93] In response, the taskforce proposed a new health structure that retained government as both the main funder and provider, but wanted the two roles separated, thus enabling, "a market to be created in which prices are set by modified competition between hospitals."[94] The creation of a new Ministry of Health dealing solely with policy advice to the minister was recommended. The reformed health system was to be funded indirectly from central government via a new National Health Commission, to six newly created Regional Health Authorities.

The emergence of a political atmosphere that expressly promoted a culture of distrust and emphasized the need to avoid both provider capture and to create mechanisms that would increase professional accountability could hardly fail to impact upon an already embattled education sector. In fact, as early as November 1985, the Education and Science Select Committee had announced its intention to conduct an inquiry into the quality of teaching on the grounds that teaching was, "the most important of all education factors," that there was clear evidence that many students were leaving school "unqualified and alienated by failure," and that there was "considerable disquiet about current methods of accountability in teaching, with many parents and communities having lost confidence in how teachers and principals were held to account for the quality of teaching."[95] Chaired by Noel Scott, the author of the 1977 report on secondary schools and now the Labour member for Tongariro, the Scott Committee began by acknowledging its debt to the many previous groups that had expressed concerns about education, including the 1987 *Curriculum Review*, and the Committee of Inquiry into Curriculum, Assessment and Qualifications in Forms 5–7. The Committee itself enjoyed a multiparty representation that included Ruth Richardson, clearly signaling that radical reforms were on the agenda. Once again the language, the concepts, and the solutions advanced by the Scott Committee were to be reiterated by the Picot Taskforce little more than a year later. Thus the Committee claimed that there was

> considerable disquiet about current methods of accountability in teaching. Many parents and communities have lost confidence in how teachers and principals are held to account for the quality of teaching. The committee believes that the key to quality depends on a proper influence for all those involved in or affected by teaching.[96]

Equity was to be ensured for all learner groups irrespective of their age, gender, age, intellectual or physical capabilities.[97] In order to achieve this goal, it was contended that children and parents had to be involved in educational decision-making, hence "the importance of a partnership between teachers and pupils and between education providers and consumers...based on full disclosure, informed discussion, and joint decision making."[98] The Committee also claimed to have "heard many critical cries from the heart with parents feeling "unwelcome in schools" and "intimidated even by physical layout of buildings." They were "hampered by lack of information about the way schools work, or how to exercise their right to influence education, felt excluded by "professional expertise" and unable to contribute from "their knowledge about their own children." They believed that they "had little or no say in decisions affecting their children's education."[99] Likewise, many community members

- expressed frustration in having grievances dealt with effectively
- felt powerless through lack of information
- overwhelmed by number of hurdles in having complaints addressed
- felt strongly that the rights of incompetent teachers are protected at the expense of pupil's rights.[100]

Several solutions were recommended. Both school and community were invited to review the "openness" of the school and its accessibility to parents and community All school staff along with the wider school community were to participate in short-listing candidates for principal positions. Regular reviews were to be tied to performance in all positions of responsibility that were to be of limited tenure. Job security was seen to be, "over-protective of teachers," and hence to work against quality.[101] It was claimed that

> professionalism requires accountability (true of medicine and law, thus also teaching). With teaching, however, unlike medicine and law, consumers generally have little choice in the teaching which their children receive. This emphasises the need for better systems of accountability in teaching.[102]

Once again foreshadowing the views of the Picot Taskforce, the Scott Committee contended that there had been an overemphasis on the Department of Education's traditional role of maintaining systems and physical provisions to the extent that its main function, the education of young people, had lost priority. Anticipating a major role of the future Ministry of Education, the Scott Report argued that the department, "must adopt a much more significant role in servicing the quality of teaching. It must more critically and constructively appraise the education system to ensure the quality of education for young people."[103]

The recommendations of the Fargher-Probine Report into continuing education and training, published just four months prior to the taskforce's first meeting were to provide a further blueprint for the language, concepts and solutions articulated in the Picot Report. In their foreword, Fargher and Probine justified taking a wider view of education than just continuing education and training, because

> significant issues include low participation rates in tertiary education in New Zealand compared with other countries; the rate at which people with technological skills are being produced; a highly centralised style of management which inhibits the adoption of an entrepreneurial approach to the delivery of services; lack of coordination in the delivery of training; and last but by no means least, the need to provide special help for socially disadvantaged groups such as young unemployed, Māori who have been disadvantaged by a Pakeha-based education system, women who wish to enter non-traditional occupations, residents of rural areas, the disabled, and the socially isolated.[104]

The report also noted; "serious lack of coordination in the delivery of education services, with decisions on what is to be taught and where shared amongst a bewildering array of organizations."[105] The Department of Education's management style was criticized for being; "highly centralized and interventionist." The proportion of resources over which institutes, college councils and principals had discretionary decision over was sometimes less than 10 percent of total expenditure which inhibited the speed at which they could respond to new situations as well as limiting their options.[106] Hence the report recommended that

> devices to improve accountability would include the negotiation of a specific charter which would contain a statement of the responsibilities, tasks and goals of each institution; the development of improved financial management systems which must be in place before institutions are given their increased delegations; the use of corporate planning techniques which emphasise the use of measurable objectives against which performance can be assessed; improved policy and planning at the centre; and more emphasis on the development of the management skills of senior staff.[107]

Finally, it should be noted that the Royal Commission on Social Policy whose views were also to profoundly influence the Picot Report, had been established in October 1986, the six commissioners being Sir Ivor Richardson, Ann Ballin, Marion Bruce, Mason Durie, Rosslyn Noonan, and Len Cook—an overall Left-of-Center grouping. The Commission subsequently developed close relationships with several government departments including education. It conducted several major rounds of travel and undertook extensive consultation around the country, including a comprehensive program of seminars, and public meetings. In addition the commission sent mail outs to some 10,000 individuals and groups, conducted statistical surveys and distributed discussion booklets. Its first meeting was held in February 1987. It was thus accepting submissions at the time the Picot Taskforce was writing its report, and was still sitting in early 1988 when the Picot Report was released. The Royal Commission was to have a direct input into the Picot Taskforce deliberations through its chair, Richardson, who was an ex-officio member.

By the time the Picot Taskforce met, all the imperatives for radical changes to the educational system were already present. The problems its members were to identify, the solutions they recommended, even the phrases in which the report would be cast, had been thoroughly rehearsed and the stage, set. The lights, as Butterworth and Butterworth have suggested, were indeed "all green" for reform.

Part 3

Elusive Consensus

Chapter 6

A Blank Page Approach

The Taskforce to Review Educational Administration was announced on July 21, 1987. It was an interesting and significant group. Chairman Brian Picot had a prominent background in the retail industry, and in the published report he was to be listed simply as a Company Director. This commercial background was to be subsequently highlighted by opponents of the reforms. In addition to being the director of several companies and a past-President of the Auckland Chamber of Commerce (1975), however, Picot also had a considerable record of public service to his credit by this time, including membership of the National Development Conference (1969) and the New Zealand Planning Council (1977–1980). He was also a University of Auckland councilor (from 1985), in addition to being the author of several papers on industrial and social matters (See chapter 3).

Picot's name had not been the only one circulating in Wellington policy circles as a possible taskforce chairman, however. As late as June 1987 the minister of finance, Roger Douglas, wrote to Marshall concerning taskforce membership. Noting that "your officials and mine have broadly agreed on terms of reference which cover the concerns of both departments," Douglas proposed that there be three people on the team—a chairperson chosen for ability to effect change and one or both the others having "some acquaintance with the educational sector but neither should be closely identified with any particular interest group. Douglas invited Marshall to consider Arthur [Doug]las Myers, a prominent Auckland business man and CEO of Lion Breweries as taskforce chair, though he conceded that Education, like Treasury, would have its own recommendations regarding the other Taskforce members."[1]

This was most certainly the case, and here at least, Education seems to have got its way. Maurice Gianotti recalls that a list of around thirty potential taskforce members was compiled within minister of education, Russell Marshall's office and it was from here too that Picot's name as a possible Chairman originated.[2] Picot's initial involvement with the taskforce thus began with a call from Marshall's office, inviting him to chair it. Picot agreed, subject to three conditions being met: being able to exercise some discretion in the choice of taskforce members, gaining a promise that the necessary funding would be forthcoming for any recommendations the taskforce might make, and an assurance that he would first have to agree with the terms of reference.[3]

Picot's discretion in selecting members, however, seems to have been limited to having some power of veto. This he appears to have exercised only once, rejecting a proposed business representative on the taskforce on the grounds that his views were too right-wing in favor of the more moderate Colin Wise, the managing director of Alliance Textiles, who had considerable educational experience as a University of Otago Council member and as a past member of a secondary school board of governors. Wise, Picot felt, could put a business point of view, allowing himself as chair to concentrate on overall strategy leadership of the taskforce.

The other members of the taskforce were drawn from diverse backgrounds but all had considerable background and experience in education. Peter Ramsay had been a former primary school teacher. Then an associate professor in the Education Department at the University of Waikato, Ramsay was also a prominent educational researcher and the author of numerous academic publications, a number of which had been highly critical of the existing education system. Margaret Rosemergy was a Senior Lecturer in early childhood and professional studies at Wellington Teachers' College. She had also been a former Lecturer in Psychology at Victoria, and Chair of Onslow College Board of Governors. Whetu Wereta was a Māori of Ngaiterangi-Ngatiranginui descent. She was also a social researcher, a member of the Royal Commission on the Electoral System, and was on the staff of the Department of Māori Affairs.[4] Maurice Gianotti was the chief executive officer for the taskforce. A senior Department of Education officer, Gianotti was at the time both a District Senior Inspector of Schools and the Acting Wellington Regional Superintendent of Education. In addition there were two full-time secretariat staff and two seconded part-time secretariat staff, as well as two wordsmiths attached to the taskforce to provide further assistance.

The full terms of reference for the taskforce reflected the vastly changed policy environment of the mid-1980s, itself a product of the intense social and economic debate that had taken place over the previous two decades. For this reason the taskforce was given a sweeping brief to examine:

- the functions of the Head Office of the Department of Education with a view to focusing them more sharply and delegating responsibilities as far as is practicable
- the work of polytechnic and community college councils, teachers college councils, secondary school boards, and school committees with a view to increasing their powers and responsibilities;
- the department's role in relation to other educational services;
- changes in the territorial organization of public education with reference to the future of education boards, other education authorities, and the regional offices of the Department of Education;
- any other aspects that warrant review.[5]

More specifically, the taskforce was charged with ensuring that the systems and structures it proposed were "flexible and responsive to changes in the educational needs of the community and the objectives of the Government."[6] The costs and benefits of its recommendations were to be identified, and were to "ensure the efficiency of any new system of educational administration that might be proposed."[7]

The Initial Meetings

Given the forthcoming General Election, the first full meeting of the taskforce on July 31, 1987 was conducted in a climate of immediate preelection uncertainty, with Picot emphasizing that if the government were to change on August 15, the taskforce's work might well be discontinued. From the outset, the timetable was formidably tight. The aim was to complete the report by the end of March 1988, although there was some allowance made for the possibility of an extension through to mid-year. The provisional timetable called for information-gathering and receipt of submissions by the end of October; the reading of issues papers and consultant reports together with the completion of face-to-face meetings with major organizations by the end of December; and the initiation of report-writing early in the New Year. There were to be seven further full-day meetings, the last scheduled for January 21/22, 1988.[8] Concerned about the tight schedule, taskforce members expressed a strong wish that all groups had sufficient time to develop and present their views to the committee, Gianotti being delegated to visit the heads of all the major organizations on behalf of the taskforce to discuss the proposed timetable.

From the outset it was agreed by the taskforce that the ensuing report would be a policy document rather than a discussion document. To this end it was assumed that debate could be facilitated by enunciating general principles that might then be commented on by interested organizations, a list of which was to be drawn up by Gianotti. In addition a number of organizations were to be invited to meet the committee, including the Māori Council, the Kōhanga Reo Trust, the Māori Women's Welfare League, the Pacific Islands Resource Centre, and Pacifika, as well as the teachers' organizations and (possibly), the Clerical Workers' Union and the Public Service Association. It was agreed that Treasury and the State Services Commission (SSC) be invited early to make suggestions to the taskforce, the chairman having invited their representatives to sit in on meetings as they saw fit.[9] A similar invitation had been issued to Sir Ivor Richardson. There were probably sound reasons for these inclusions. Picot might well have felt that Treasury and SSC representation would avoid a situation whereby the taskforce might reach a consensus but subsequently have to reargue its case with the two departments. Moreover, in seeking a closer relationship with Richardson, Picot had possibly sought to alleviate some concern amongst Royal Commission members that the setting up of committees to deal with both education and health cut across their own brief.[10]

The material distributed to taskforce members at the initial meeting included the attached recommendations of the Nordmeyer EDC Working Party, and the Fargher-Probine Report, both of which had supported educational devolution.[11] Given the tenor of these documents and the general policy climate, it is hardly surprising that the minutes to the first two meetings confirm the rapidity by which taskforce members collectively resolved on radical change rather than simply tinkering with the existing system, a decision that seems to have surprised even Smelt. As Picot pointed out to taskforce members, however, "the presence of a serious disease is the justification for radical surgery."[12] Accordingly

> it was agreed that the committee's reports would promote the proposals believed to be logical and sustainable without regard to the acceptability of those proposals to particular interest groups. The starting point of committee deliberations would be a "blank page" approach on which an efficient and effective administrative structure can be erected.[13]

In a later interview, Picot was to recall that the "blank page" approach was taken because the difficulties inherent in trying to massage an outdated and inappropriate 110 year-old system into shape were simply too great to meet present and future demands. He rejected as "absolute rubbish" any suggestion that the taskforce was simply trying to save money or to create a competitive education system.[14] What does appear to have more influential were the poor exit qualifications in some secondary schools, particularly among Māori students. The 1986 statistics the taskforce had access to, for instance, revealed that while only a single secondary school on Auckland city's wealthy North Shore had more than 25 percent of students leaving without a single School Certificate subject, in South Auckland with its high Māori and Pasifika population, it was true for no fewer than 13 out of 16 secondary schools.[15]

Picot in particular, had also reportedly been outraged about the degree to which the existing education system penalized outstanding teachers while sheltering the less competent—a factor already noted by the Scott Report. Gianotti recalls that:

> We went to a little school... on the outskirts of Carterton where people, because of the young teacher who was there, because he was very good, people were bringing their kids from the far side of Carterton, carpooling... the school had gone to a three-teacher. It now had to be advertised and the job was filled by somebody from a 4 or 5 teacher school where the roll had dropped and was now entitled to a 3-teacher. The fellow... who has built it up had to go.[16]

Even at these early meetings, there was an acute awareness amongst taskforce members concerning the general direction of existing critical comment on the system. The recommendations of the Nordmeyer Committee appended to the agenda for the initial taskforce meeting were especially significant in this context. Recommendation 8.10 had argued that secondary school Boards of Governors should retain power to appoint staff. Recommendation 8.11 had suggested that in the primary sector, each school committee should more closely participate in the appointment of staff and, in the case of the principal, should through its chairman have a voice wherever possible on the board's appointments committee.[17] Recommendation 8.49 had advocated that the department's close control over specific sectors of the education system be reduced by transferring wherever possible the power of decision making to local or district institutions.[18]

The Fargher-Probine Report was also discussed at this initial taskforce meeting. Picot stated that he had liked the general thrust of its proposals, and he emphasized that the interface of secondary schooling with technical and continuing education was an issue for the taskforce to consider.[19] One of the tasks allocated to the Secretariat at the conclusion of this meeting was to make up sets of all the papers that members had indicated a wish to study more closely.[20] Thus, the collective impact of the Nordmeyer Committee and the Fargher-Probine Report as well as a raft of previous sector reform documents can be clearly discerned in the taskforce's early consensus that the system was "too complex and too unresponsive."[21]

At this initial meeting taskforce discussion focused on the role of the school principal and the centrality of quality control within the educational system. It was agreed that any proposed structure had to include accountability for outcomes, at every level of system, and that whatever was proposed was to lead to "excellence of outcomes," although it was conceded that these might be different for each taskforce member.[22] Members were initially invited to make their own positions clear. Wise emphasized the need for the schools to produce "people who could contribute to the economic growth of the country but at the same time are tolerant, caring and cooperative."[23] Picot thought that the so-called hard subjects might have been downplayed in education, and that "a balance had been lost and needed redressing."[24] Wereta believed that the definition of an educated Māori person was one who was "truly bi-cultural, who knows their Whakapapa and who can function effectively in both cultures." While Wereta did not expect schools to teach Māori language and culture, she thought that they should be supportive of them in the same way as they were of European culture. Like the other taskforce members, she emphasized devolution, pointing out that both the Health Department and the Department of Māori Affairs were already promoting it.[25]

Even at the initial meeting, therefore, devolution was clearly identified in the initial meeting as being the solution to many perceived systemic and educational problems. Discussion accordingly focused, not on the advisability or otherwise of devolution, but on how this goal could be best achieved, what safeguards would be required, and how it could be made most effective. This decision in favor of devolution was to raise a number of concerns. Picot articulated a misgiving that was to surface on a number of occasions both prior to and after the release of the final document, namely "that devolution of power might lead to a 'hijacking' of some institutions by a particular group who have a strong point of view and are organized enough to gain control democratically."[26] Lange himself was to make just this point on several occasions over the weeks that followed.

Given that an early intention of the taskforce was to recommend that decision making be devolved as far as was practical to school level, the role and quality of principals and vice-principals was quickly identified as being of critical importance. There was also some preliminary discussion on the role and function of a central ministry of education, it being agreed that the new ministry would assume national functions that could not be undertaken by devolved authorities. These included the setting of minimum standards and provisions, the development of national curriculum guidelines, and performing accountability and efficiency checks.[27]

That the taskforce had already decided on radical reform without much Treasury and SSC prompting is clear from the agreement reached by the conclusion of this initial late July meeting that

1. A structure must be designed which promotes excellence, standards of attainment and performance by all pupils.
2. The system must be flexible and responsible to consumer demand, that is, it must be "user-friendly."
3. Decisions on resource allocation should be placed at the school or, as close as possible to the school and only when there are quite compelling reasons should decisions be made beyond the school.
4. The number of layers of decision making should be kept as few as possible.
5. The number of sectors should be limited to the minimum number for efficient working. If district authorities are necessary they should cover early childhood, primary and post-primary schooling.
6. School managing bodies and district authorities, if they are required, should have the maximum possible authority to carry out their functions. They should be given cash grants and required only to provide minimum levels and standards of, for example, staffing, accommodation, salaries and equipment.
7. Any district authorities should be of the minimum size necessary to enable them to have the staff equipment and facilities to carry out their functions.
8. The administrative structure of the system should be as "lean" and efficient as possible.[28]

The second meeting of the taskforce, on August 17, examined more closely the possible advantages and disadvantages of devolving authority. In many ways, the ensuing discussion centered on the existing tensions between central and local control that had been so astutely anticipated by Leicester Webb in advocating further devolution almost exactly fifty years before.[29] For instance, taskforce members quickly identified the competing interests likely to be present in any devolved system. It was agreed that the devolution of authority and functions could make the system more flexible and responsive, but that there would also have to be controls specifically designed to overcome "capricious or arbitrary action" on the part of newly created bodies. There was, nevertheless, a strong feeling that opportunities for choice be maximized for all groups—for example cultural groups or those with particular beliefs who held strong views about what schools should provide.[30] This latter conviction also reflected an awareness of both Māori demands for devolution, and the expansion of Te Kohanga Reo.

Given the debate that had accompanied the two previous reviews of the curriculum in 1984 and 1987, it is noteworthy that the taskforce considered a national curriculum to be essential. The section entitled "Aspects of Learning," in the 1987 *Curriculum Review* was seen as a good starting point for discussion, although the taskforce accepted that an inevitable outcome of devolving power and responsibility would be increased diversity. Hence, the new system was to be designed to encourage a high standard of educational outcomes for all groups.[31] In turn this view led the taskforce to a discussion of the government's proper role in education and to a central paradox of devolution both pre and post-reforms that has in fact remained unresolved to this day—namely, the more a system is decentralized, the more it is deemed necessary to have a centralized audit for accountability purposes.[32]

Finally, there was some discussion amongst taskforce members regarding the status of Treasury representative, Simon Smelt and SSC representative, Marijke Robinson, centering on precisely what their status during taskforce sessions was to be. It was decided that they would be regarded as part of the Secretariat and as individual appointments, attending with the invitation of the taskforce and acting as a resource for the committee. They were not, however, to enjoy voting rights.[33]

A one-page document signed by Picot and dated August 31, 1987, recorded that the first two meetings of the taskforce, while somewhat unstructured, had nevertheless established a timetable and agreed operating procedures, made contact with key organizations who were requested to make early submissions, sent letters to informal organizations such as tribal authorities, received papers from Treasury and SSC, and circulated background papers on the existing education system. It was also once again asserted that the taskforce had opted for a "blank page approach...rather than try and massage existing structures into a more appropriate form."[34]

Policy Precedents for the Taskforce

With this account of the first two meetings in mind, it is appropriate at this juncture that we now turn to some of the papers available to the taskforce as they deliberated during the second half of 1987. Of the considerable volume of documents available to the taskforce during its deliberations, the Treasury Briefing Papers were to become the most controversial, with education policy commentators in particular subsequently singling them out as the single most important factor in determining the shape and direction of both the Picot Report and *Tomorrow's Schools*. Entitled *Government Management* and released in late August, the Treasury Briefing Papers actually consisted of two substantial volumes. The 471 page first volume dealt with the broader issues that would face the incoming government after the forthcoming general election.[35] It was observed that although the economy had grown stronger than expected, high inflation coupled with a gloomy outlook for international trade threatened to blunt the gains from the market liberalization introduced under Labour.[36] Although not directly concerned with education, this volume also included a brief education section, pointing out that the state education system was one of the nation's largest enterprises with an annual expenditure of approximately $3 billion. Accordingly, government intervention in education was seen to raise four significant issues: who pays; who chooses; who benefits; and who is accountable.[37]

These issues were taken up in more detail in the second volume, prepared by Simon Smelt and Michael Irwin. This volume was entirely devoted to education, justified by its authors because of, "the complexity of the issues involved (there are no easy answers), the importance of the educational sector to wider issues of social equity and economic efficiency, and the extent of apparent public concern about the public education system." Although Treasury viewed some recent changes in a positive light, it warned that "substantial elements of current government expenditure are, at best, ineffective when viewed in terms of the equity and efficiency concerns that justify such expenditure."[38]

Like the taskforce itself, the Treasury analysis raised issues that Webb writing in 1937 would have clearly recognized, but with a new twist that owed much to the debates of the 1970s and 1980s. It was pointed out that New Zealand had an overwhelmingly state education system with a small private sector and hence little consumer choice, apart from the "vigorous and fast growing Kohanga Reo movement for preschoolers." Although the dominant form of state intervention was direct provision of educational services, however, Treasury observed there was no *a priori* reason why intervention should take this form. Here Treasury noted the recent Access program which subsidized training providers who could be private, likening this initiative to Te Kohanga Reo, which provided a separate Māori delivery mechanism in response to call for Māori control over resources. Although government educational services were provided virtually free to consumers and parents through to postgraduate level, there was little targeting of assistance for disadvantaged groups, and a high degree of centralized educational control that produced rigidity and a slowness to react to changing demands. Hence, according to Treasury, Te Kohanga Reo and Access could "be seen as ways of bypassing a system that that has failed to react sufficiently, or fast enough, to the needs of specific groups." Perhaps picking up on the anti-provider-capture recommendations of the Cartwright Inquiry and the Scott Report, Treasury asserted that a unified teacher workforce hindered the application of incentives for high performance, and discouraged sanctions for poor performance.[39]

Noting that secondary schools had received the most public criticism, Treasury argued that PAT tests demonstrated statistically significant declines in attainment levels in listening comprehension, mathematics, and reading from late 1960s through into 1980s. Worse, the education system as a whole "did not appear to have succeeded in redistributing opportunities towards least able." These failures were reflected in a growing Māori/non-Māori achievement gap, high youth unemployment, youth crime, delinquency, and vandalism. Treasury thus concluded that, "as far as the limited evidence admits, the quality of formal educational performance at secondary level appears to have been declining, whilst the real unit cost of financial inputs from the Government has been rising."[40] Hence, its recommendations for secondary schools focused on the long term benefits of education through the core curriculum, coupled with the imposition of minimum standards. There was to be an increased flow of information to educational consumers and a maximization of user choice, while the core curriculum and minimum standards were to be retained. Targeted funding for disadvantaged groups and incentives for good teacher performance, together with improved inspection and more accountability for secondary schools, were also advocated.[41]

Chapter 8 was particularly noteworthy in that it focused specifically on the causes of Māori underachievement. Here, Treasury was able to cite a wealth of recent academic research both in New Zealand and elsewhere, for both its diagnosis and its remedy. This included Harker's work on social control, hegemony and cultural capital, Walker's and the Smiths on Pakeha capture of Taha Māori and the struggles of urban Māori, Judith Simon's revelations of how teacher's practices contributed to the denigration of Māori culture in classrooms, Benton's advocacy of a separate Māori-controlled system, and the research of Nash on educational inequality. Treasury was also able to point to increasingly strident calls for education to be held to account for failing to acknowledge the primacy of partnership and the special place of Māori under the Treaty of Waitangi, as exemplified in the 1982 report, *Race Against Time*, and the conclusions of the Waitangi Tribunal. It backed these imperatives with numerous citations of overseas research by Bernstein on language codes, Bourdieu on habitus, and Boudon on secondary effects. Smelt was later to claim that it was research evidence such as this rather than the example of Thatcher's government in the United Kingdom, which had largely fuelled Treasury present concern.[42] Accordingly, Treasury envisaged policy initiatives as proceeding on three broad fronts. Māori required continuing assistance in revitalizing their language and culture through Te Kōhanga Reo, while mainstream secondary schools were to offer more bilingual education. Pakeha knowledge of, and attitudes toward, Māori were to be addressed, although Treasury warned that this could easily lead to a Pakeha takeover of Māori culture, as had already occurred with Taha Māori. Most important, the best way of ensuring that Māori could gain the knowledge that would enable them to succeed in modern society had to be found.[43] This latter goal meant that there was a need for "specific institutional devices to ensure that Māori aspirations are not filtered out by inappropriate Pakeha institutional frameworks." There was also a need for a "Māori component that, with its emphasis on community, acts as a strong counterweight to the individualism of the Pakeha."[44] As we have seen in chapter four, this last recommendation had its antecedents in secondary school criticism dating back to at least J.H. Murdoch's 1944 assertion that "the strongly developed community life of the Māori is a necessary corrective to the selfish individualism that is so marked a feature of Pakeha society."[45]

Even in its most controversial argument that education was not a "public good," but instead benefited individuals, Treasury was able to draw on longstanding Left-liberal and conservative concerns that, although education was funded by the community, those who provided educational services frequently sought to defend and develop their own interests. Thus, in critically examining the four key educational issues of who paid, who chose, who benefited, and who was accountable, Treasury effectively summed up the widespread and ideologically diverse dissatisfaction with state secondary education that had increasingly marked the previous forty years. Despite the claims of liberal-progressive educators since Beeby's time, secondary schools had neither produced equity nor equality of outcome (radical sociologists, revisionist historians, feminists). Deeply embedded structural and cultural inequalities meant that not all families were able to purchase or take advantage of education, with Māori in particular bearing the brunt of systemic failure (NACME; Turangawawae Marae [1984]; *Taha Maori and Change* [1986]. Institutions were unresponsive to the needs of consumers (CPA; CARE, *Race Against Time*, the Benton research, radical Māori agitation, the Gibbs Report). Middle class providers, especially teachers and Pakeha bureaucrats had effectively captured the system for their own ends, with consumers often unable to hold them to satisfactory account (the Erebus Inquiry, the Scott Report, the Cartwright Report).[46] Solutions, however, could be glimpsed in promising new initiatives toward devolution (CEIS, Te Kohanga Reo, Access).

In contrast to Treasury, although the Department of Education was aware of at least some of the critiques of education that were circulating during the first half of the 1980s, its major spokespeople do not appear to have grasped the degree to which the major concepts underlying devolution had become interwoven into a more complex policy discourse that drew upon a mixture of radical ideologies, rather than an exclusively neoliberal discourse of privatization and consumer rights. Consequently, the department underestimated both the power of the new rhetoric and its degree of popular appeal across the community. An undated Draft Departmental submission to the taskforce prepared for the late October meeting of the taskforce is symptomatic of the degree to which the department had lost touch with the grim new mood for change. The submission warned that all attempts to reform the education system had ended in failure.[47] It pointed out that

> the privatisation of education is not in our view an acceptable or a realistic way of arranging for the provision of one of the vital public goods of any modern society. Arguments for consumer sovereignty are symptoms of a sense of dissatisfaction that some members of the public have about public education. They do not offer the basis for a solution to the concerns they reflect.[48]

This view was but a reiteration of the department's stance through much of the twentieth century. It was at least conceded any reformed education system would at each level have clear-cut responsibilities and accountabilities, with no overlaps or gaps. Furthermore, the department claimed, "the central task of a reorganized Department of Education—one which, in our view, would be more appropriately called a Ministry of Education—would be to provide disinterested policy advice to Ministers."[49]

Unfortunately for the department, however, these recommendations now offered a palliative in the face of the considerably more radical solutions now being openly predicted in the national press by prime minister and minister of education, David Lange. Picot in a later interview claimed that Lange did not attempt to influence the views of the taskforce, and that the only concession made to the prime minister was that he was to receive an advance copy of the final report prior to its being published.[50] This is supported by Gianotti, who later recalled that Lange had little direct impact on the taskforce's deliberations. Picot himself had initiated fortnightly meetings with Annette Dixon and Anne Meade in the Prime Minister's Office to discuss general progress, but there was no interference.[51]

Indirect pressure, however, was another matter entirely. As we have already seen, Lange was already going public in expressing his personal desire for radical reform. In early November, a front page leader in the *Evening Post* featured a speech delivered by Lange at the end of a Labour Party social debate earlier that day to the effect that the education system was long overdue for reform. The article cited the prime minister's opinion that

> the Department of Education is basically an educational institution. Its interest as an educator colours the advice it gives to the Government. If the Government asks the department for advice about funding, it is like asking a child how much it wants for pocket money.[52]

According to the *Post*, Lange had further intimated that he intended to dismantle the Education Department and, in the now familiar pattern of public sector reform, to establish a Ministry of Education separate from the organizations which actually offer public education. Henceforth the new entity's tasks would be to conduct research, to advise the government, to set standards for educational institutions, to make sure these standards were met, and to provide the necessary funding. In addition, it would be charged with working out measures that would satisfy the public that those teachers who could not teach would find other occupations while those who excelled would be encouraged.[53]

Moreover, by early November the Picot Taskforce had received a series of short but pointed Treasury papers outlining the details of radical reform. A four page document listed what was described as a number of efficiency considerations, including the problems of government intervention, central control, management, and accountability.[54] A rather more specific Treasury paper included an appendix significantly labeled "The Education Industry: a possible model."[55] In addition a paper by Smelt, dated 9 November, discussed the question of educational accountability.[56]

The taskforce had also received copies of a report it had specifically commissioned to assess the functions of the existing Department of Education, prepared by W.D. Scott Deliotte Ltd, a private Wellington-based management consultancy firm.[57] This report actually consisted of a series of working papers designed to "assist the members of the taskforce to understand what the department really is, and what it really does."[58] Several underlying problems were highlighted.

- The stress is on action, not results.
- The absolute priority is on answering ministerial questions.

- There is little real accountability for results.
- Most managers have little control over most of their resources (e.g., Staff).
- Divisions are grouped so that several have to be involved in producing a result.
- Overlaps and inefficiencies exist both between Divisions and Regions, and between the department and boards.
- The decision-making process throughout the department has progressively become excessively consultative.
- The whole system of management and administration is bureaucratic, with most of the associated inefficiencies and reduced effectiveness.[59]

The report's conclusion was thus highly critical of the department's organization and performance. It was alleged that there was no responsibility for budget setting or expenditure control. Education was described as "the product," for which the department had no line responsibility either for delivery or quality control. Few senior managers were seen to have any grasp of management concepts. The department's culture was stigmatized as "reactive and stultifying."[60]

Finally, by late November an extensive list of abstracts featuring no fewer than 131 relevant papers and commentaries generally supporting radical educational change was available to taskforce members. These were drawn mainly from academic and professional publications, books, magazine articles, parliamentary debates, and newspaper features from Australian, North America and the United Kingdom as well as New Zealand.[61] It is noteworthy that the voucher issue featured strongly in the list, with the first 17 items dealing either wholly or in part with educational vouchers. Of these a paper entitled "Education Vouchers" dated February 2, 1987, produced by the New Zealand Government Research Unit, is particularly revealing. This paper examines in detail the various ramifications of what had by then become a global issue. The critical research of New Zealand academics such as Snook, as well as the more supportive views of English neoliberal voucher lobbyists such as Marjorie Seldon of Friends of the Education Voucher, and Norman Macrae, writing for the *Economist,* was cited to illustrate both sides of what was acknowledged to be a complex debate. The paper went on to point out, however, that while there were no explicit tests of the voucher concept immediately available, the Youth Training Scheme in Britain and the introduction of Te Kohanga Reo in New Zealand permitted some useful conclusions to be drawn. Turning to Te Kohanga Reo, the paper emphasized that the scheme had emerged as a Māori educational initiative against initial objections from the Department of Education to the extent that "consumers were thus able to direct funding towards providers by their decision to use those providers." The result was that

> Te Kohanga Reo has now become an accepted alternative in the pre-school area and its funding has shifted from the ad hoc use of employment programmes to specific provision through the Māori Affairs Department. It remains, however, as evidence that, provided they have access to financial resources, and the freedom to design or demand programmes which they believe meet their needs, then even disadvantaged groups are capable of putting very effective programmes into action.[62]

Despite these policy precedents, the voucher scheme as such appears not to have been seriously entertained by the Picot Committee. However, the Te Kohanga Reo and Kura Kaupapa examples were to provide further justification for the committee's decision to provide a mechanism through which parents could choose to either withdraw their children from mainstream schools, create schools within existing schools, set up their own schools, or to homeschool.

The voucher debate in New Zealand hinged on issues of choice related particularly to questions of equity and fairness. One indication of the influence on taskforce deliberations of contemporary critiques concerning the fairness of New Zealand education, particularly the adequacy of a single system to accommodate diversity can be found in the numerous references to the Royal Commission on Social Policy in surviving documents. The initial Report of the New Zealand Council for Educational Research published in October 1987, and the final report of NZCER published in November 1987 are of particular interest, with the latter in particular emerging as an extensively researched two-volume document commissioned by the Royal Commission on Social Policy.

Given the whole tenor of Māori advocacy over the previous decade, it was perhaps predictable that the first paragraph of the initial volume of the latter report set the tone for what was a strongly worded critical analysis of New Zealand's educational achievements over the first one hundred years of its existence. As with the many previous critiques of education furnished not only by radical Māori advocacy groups, but by many other critics over the previous two decades whose common conviction that the system had failed effectively masked their ideological differences, this was an analysis based not so much on the notion of equality of access, but rather on the perceived unequal outcomes of those groups identified as being disadvantaged:

> This report examines the fairness of the New Zealand education system in relation to seven groups identified as likely to be disadvantaged. They are: those from low socio-economic status homes; girls and women; Māori, Pacific Island groups; the disabled; ethnic migrant groups; and rural dwellers. *Allowing for overlap between these categories, this accounts in fact for at least three fifths of the population included broadly within the education system* (italics mine).[63]

The conclusion reached by what was highlighted as the "rather depressing evidence reported here," was that "the situation of disadvantaged groups will only improve if there are major changes in our education system."[64]

The pessimism of the first volume of the NZCER report was amplified and extended in the second volume, written by NZCER's researcher on Māori education, Richard Benton.[65] Benton believed strongly that the high failure rate of Māori children in an essentially Pakeha state education system clearly demonstrated that the cultural and social needs of Māori were not being met. The system was essentially closed and self-perpetuating. In addition, he claimed that there was "widespread research evidence that teachers as a group assume that Māori children at all levels are likely failures."[66] Drawing upon Karl Polanyi's 1944 analysis of the impact of market economies on societies and citing the disappearance of Māori schools in the 1960s as a prime example, Benton argued that a closed educational market was "essentially totalitarian in nature, absorbing, and where possible destroying, alternative structures, even alternatives which it has generated itself."[67] In some respects this was essentially similar to what many Left-liberal and neoliberal critics had long been arguing. Benton's conclusion, however, and particularly the ideology that underpinned it, are particularly revealing

> According to Polanyi's analysis, the community can recreate itself by a process of social restoration involving what would in Māori terms be called mana motuhake. This is just what the kohanga reo and many other Māori social initiatives are aimed at. It cannot be said, therefore, that Māori people must inevitably be victims in the market place. On the contrary, they may well be able to exploit some of the new market forces to their own educational and social advantage, and use this advantage to defend themselves and their institutions from further erosion. *It is because of this that some of the ideas contained in the 1987 Treasury analysis of the management of the education system (1987) are, in my opinion, to be welcomed* (Italics mine).[68]

Polyani's attractiveness to an Antipodean policy environment that had become so strongly iconoclastic and eclectic as New Zealand's during the

1980s can perhaps best be understood in the context of his highly-individualistic and radical approach to societal reform. Born in Vienna in 1886, Polanyi had been educated as a philosopher before fleeing Central Europe first to Britain and then to the United States. His brother Michael Polanyi, who like Karl was to gain an international scholarly reputation, had been closer to his near-contemporary Friedrich Hayek in his belief that the state could no longer effectively run a modern economy, Karl Polyani, however, sharply differed from both men in placing little faith in any of the West's major political, social and economic ideals; hence his attractiveness to many disillusioned academics, post-1968. In his-best known work, *The Great Transformation* (1944), Karl Polanyi emphatically rejected the free market system along with the various attempts to regulate or militate it, including socialism, liberalism, and fascism. Instead, he claimed to have rediscovered what he termed "the primacy of society."[69] In advocating the shifting of Western industrial civilization onto a new non-market base once the Second World War ended, Polanyi drew upon a highly idealized view of non-Western tribal societies for his ideas of a model self-governing and independent community—hence he particularly appealed to those who advocated Māori self-determination in education during the mid-1980s. Polanyi's eccentricities, however, have been critiqued in a recent book by Roger Sandall who describes how Polyani's delusions of romantic primitivism, and suspicion of Western-style democracies led him into an, "obsessive search for an idealized pre-capitalist culture" that never actually existed.[70]

Soon after the implementation of the reforms, a perceptive paper by Logan Moss, a Waikato University education lecturer, maintained that, in its frequent use of the word "community" the Picot Taskforce was consciously reinvoking a legend common to virtually all European societies: that of an idealized collection of small, cohesive communities beset, torn apart, and finally reduced to servitude by a malevolent outside force.[71] Certainly, as we have seen, Picot and a number of other critics of New Zealand society in the late 1970s and early 1980s held an essentially similar view. Moreover, given that the school is one of the principal institutions through which any community ensures its perpetuation, it may well have been true that any indication that an outside force (such as an education department) had captured the schools for its own ends was likely to be interpreted as a threat to the organic integrity of communities. In this situation any proposal for radical education reform had to be seen to deal with the central issue of freeing communities from the yoke of the system.[72]

Be this as it may, Benton's hopes for a fully devolved, Māori-controlled education system resonated with those of Treasury and SSC, even though they proceeded from seemingly different ideological positions. Thus while Benton followed critical Left-liberal New Zealand academic commentators such as Boston and Snook in rejecting a voucher system as proposed in the National Party's 1987 election policy, he nonetheless favored "a freer, more flexible, decentralized and innovative approach to education" that an open market might convey, particularly in giving Māori families the power to take control of their children's education.

Parental choice then was, predictably enough, a major theme in the list of abstracts presented to taskforce members. A number of abstracts were directly related to Rogernomics, and particularly the issue of privatization in education. An abstract of an article in the *New Zealand Listener* of July 11, 1987 by Shane Cave cited business leader Hugh Fletcher's view that a big omission in the reform process to date had been education. Although the state should provide families with the resources to pay for education, schools, Fletcher argued, parents should have the right to choose their preferred school. Another abstract outlined a study by Richard Elmore of Michigan State University. Writing for the Rand Corporation, Elsmore believed that there were good grounds for increasing client choice in education, while avoiding the extremes of centralization or devolution.

Fortuitously for the taskforce, similar reform models existed in a number of Australian states, and these were the subject of a number of abstracts, presumably included to provide the committee with some concrete examples of what a reformed education system might look like. The Victoria Ministry of Education's Structures Project Team's series of booklets produced in 1986–1987, which set out the proposals of the Australian project team, were described as

> giving a brief rather similar to our taskforce. While the Victoria State circumstances do not precisely match those of New Zealand, the proposals have a similar objective, "self governing schools" and the proposals will be of interest to Taskforce members.[73]

In fact, the taskforce was to find the Victorian reforms to be of sufficient interest to commission a more substantive report from SSC (discussed below).

In view of later debate over the nature and extent of Treasury and SSC input into the taskforce it should be noted that there are some indications that SSC and Treasury pressure, far from unduly influencing the report writing process, may have served to arouse the ire of at least some taskforce members. Both Robinson and Smelt reportedly had considerable input into taskforce discussions, with the former being sufficiently vocal in her opinions for some taskforce members to demand that she be considerably toned down.[74] Reporting to the Prime Minister's Office in February 1988, Annette Dixon astutely observed that

> the Taskforce met for three days last week and have made substantial progress. While there are still major areas of disagreement (mainly between SSC and Treasury and the rest of the Taskforce members), they have agreement on the issues relating to the administration of schools. The contentious areas relate to the policy/advisory/implementation functions at a national level. It is possible that there may be a minority report, although the group has so far been able to accommodate most of the concerns. It is possible that they may report 1–2 weeks late.[75]

In the event the Treasury and SSC representatives were excluded from the final meeting of the Taskforce and it has been claimed that the final drafting saw some of the recommendations they had strongly supported being effectively reversed.[76]

During the latter stages of their deliberations, the receipt of two further reports, probably served to convince the taskforce that their own document was at least proceeding along the right lines. In late February the taskforce received an advance copy of the Watts Report critique, written by SSC Commissioner and Deputy Chairperson, Margaret Bazley. The Watts Report, sought to defend a centralized, publicly funded system of universities. Emphatically rejecting calls to reduce public funding and to introduce market forces between tertiary providers, the Watts Committee instead requested a significant increase in funding to meet sharply rising demand, and proposed to raise efficiency by increasing staff remuneration and reforming internal management practices.

In her critique, however, Bazley implied that the Watts Report's evaluation of university education was too superficial as well as being rooted in a male-dominated past. As a result its attempt to justify the retention of the University Grants Committee reflected "an ethos of splendid isolationism." It employed "a doubtful analysis and logic," and "a selective use of evidence," while failing to develop "a systematic educational philosophy" that straddles the philosophical, strategic and political aspects of policymaking.[77] In phrases that neatly encapsulated the prevailing policy ethos in regard to social services including education, Bazley observed that

> the government is also committed to rationalisation of current institutions and structures. It is determined to reduce its involvement in service delivery, to off-set provider-capture with greater consumer sovereignty, and to use market mechanisms to raise efficiency in its social delivery systems where appropriate. It is seeking to establish new accountability structures through explicit contracts between funders and providers.[78]

Early in March 1988, the SSC Report on the Victorian educational reforms was duly received by the taskforce.[79] Written by Dr R.J.S. Macpherson, the Senior Management Officer for SSC, the report detailed three major reforms in Victorian education since 1979, along with the lessons these provided for New Zealand. The report contended that in May 1979 the then Liberal minister of education, A.J. Hunt and his assistant minister, Norman Lacey, in reviewing education for the newly elected Liberal government, had discovered "an over-elaborate organization virtually paralyzed with structural conflict."[80] Macpherson claimed that while Hunt and Lacey had acted with determination, they had faced determined opposition from those with vested interests in the status quo. Although the Director-General had only moved cautiously to devolve powers

> the most vehement opposition to the proposed devolution of personnel functions came from the primary school teachers union which claimed low management readiness, especially in many isolated small schools. Much of the rhetoric featured self-interested centralism.[81]

Macpherson conceded that teachers were jubilant when Labor was elected in April 1982, with half of its new members being teachers. Moreover, Hunt's successor, Robert Fordham, soon created the new post of Executive Director (Schools), giving substantial policy advisory powers to the newly created State Board of Education, and emphasizing the need for collaborative decision-making processes. Macpherson claimed, however, that these moves had only served to stall genuine devolution, while regional structures were expected both to provide school support services and to act as a buffer between localism and centralism.[82] By contrast he observed that Fordham's successor, Ian Cathie, had decided that a genuine transfer of power from "the Centre" to school councils and managers was necessary to make schools self-governing. The recommended measures included the promulgation of state-wide curriculum guidelines, staff selection, employment and promotion at school level, bulk funding and governance and management/governance training of school councils and principals.[83] Macpherson reiterated that devolution had been "bitterly opposed" by the teachers' unions, but he nevertheless recommended that the Picot Committee note his report, and also call for a further report on Victoria's post-compulsory education reforms.[84] It would be interesting to know if this latter report, if actually written, eventually found its way to the Hawke Committee on Post-Compulsory Education and Training that was to meet later that same year.

The Picot Report was finally released in April 1988. Given the tenor of deliberations and the widespread dissatisfaction with existing educational structures and outcomes across virtually the entire spectrum of New Zealand society that had been reflected in the intense policy debates of the previous two decades, it was hardly surprising that the Report proposed an entirely new educational structure which aimed at a simplified administration where decisions should be made "as close as possible to where they are carried out."[85] Emphatically rejecting recommending minor changes or adjustments to the existing structure, the committee emphasized that "the comments made to us in submissions and our own inquiries convinced us that the time has come for radical change in our education systems' structure."[86] These had complained of "an overly high degree of centralization in our education system," to the effect that "virtually all power and decision making comes from the centre."[87] The result was wasteful duplication, problems of information sharing, slow responses to needs, and vulnerability to "the influence of pressure group politics." Moreover, professional and bureaucratic interests had tended to predominate over the interests of educational consumers, particularly minority groups.[88]

Perhaps the key Picot Report recommendation was that individual learning institutions were to constitute the basic unit of educational administration, because these were where there was "the strongest direct interest in the educational outcomes and the best information about local circumstances." The running of individual learning institutions was to be a "partnership between the teaching staff (the professionals), and the community."[89] However, as the state provided the funds and retained a strong interest in educational outcomes, there were to be national objectives and clear lines of responsibility. This was to be reflected in the charter which, as the contract between institution, community, and state, was to be the "lynchpin" of the entire new structure. Whetu Wereta later claimed that the charter notion came from Peter Ramsay.[90] As we have seen, however, the concept was not unknown in educational policy circles by this time. In the new arrangements recommended by the taskforce, every institution was to be held accountable for meeting its charter objectives by an independent Review and Audit Agency reporting directly to the minister of education.[91]

The Department of Education was to be abolished and in its place was to be a new Ministry of Education. This was to provide policy advice to the minister through an Education Policy Council as well as implementing national education policies approved by the minister of education. The numerous critiques centered on provider capture that had been an increasing feature of the last ten years in particular were reflected in a much stricter demarcation of future bureaucratic roles and functions. In order to avoid possible confusion of roles and to keep policy free of what was seen as the self-interest of educational bureaucracies, the new ministry was to have no part in the provision of services. It was, however, to provide policy advice to the minister through an Education Policy Council consisting of four senior ministry officials and four external appointees.

Education boards were likewise to be abolished, leaving learning institutions free to choose their own services.[92] For decades, education boards had been the subjects of controversy, with successive calls for their abolition coming to nothing. A number of ministers of education had criticized them for blocking their own policy initiatives and Russell Marshall himself appears to have had experienced periodically difficult relationships with education boards. Their removal, therefore, was no real surprise. Gianotti, however, has recalled that Picot had always been concerned about removing the additional layer of administration represented by the education boards, but eventually conceded its logic, given that what the taskforce was recommending effectively left them with nothing substantive to do.[93] Later, Picot was to express his regret that the skills of former education board employees were largely to be lost to the education system. The taskforce had, in fact, hoped that an urgent working party would be set up to find ways of marketing their skills but this was never done. "If I had a chance to do it again that would be in blazing capitals."[94]

Community education forums were to be established to promote debate on educational issues and concerns. To safeguard the interests of parents in a situation where education providers were likely to remain better funded than educational consumers, a Parent Advocacy Council was to be established. This body was also envisaged as furnishing advice to parents who sought programs for their children which were compatible with the national objectives, but which no school was able to provide. When at least 21 children were involved, provision existed for the Parent Advocacy Council to negotiate with the Ministry to set up an alternative institution which would have access to state funding, "perhaps as a 'school within a school.'"[95]

The dominant role biculturalism had gained within the policy environment as a result of Māori agitation, supported by an increasing number of Pakeha educators was inevitably reflected in the Picot Report. As far as Māori were concerned, the Report asserted that

> the Māori people have told us that they want their children to be bilingual and bicultural, and at ease in both the Māori and the Pakeha worlds. As well, they want the opportunity for all Māori children to be educated in the Māori language, in an environment that reflects Māori values and uses Māori forms. We believe our structures will help achieve these aims.[96]

The Picot Report's recommendations for improved coordination between primary and secondary institutions were an acknowledgement of the concerns expressed by Left-liberal academics, teachers' unions, business organizations, politicians, conservatives, and neoliberals about the inadequacies of existing assessment regimes and the consequent wastage of young people throughout previous postwar decades, culminating in the Probine-Fargher Report. It was conceded that "too many people leave school disaffected and with no formal qualification." To remedy this situation, the Picot Report proposed the establishment of a single state authority which would validate all non-degree courses. The new authority would be responsible for developing a single coordinated system of course offerings and credits to be available full-time or part-time, from any learning institution.[97] Efficient management of the education portfolio was now to be the maxim, and this required a comprehensive property management plan.

Given the considerable number of often conflicting articles and reports received by the Picot Committee on the controversial issue of choice, the Picot Taskforce opted not to make specific recommendations. It did, however, consider that existing zoning of secondary schools had the effect of diminishing secondary school autonomy.[98] Hence the report stated that

> the exercise of choice will be enhanced if zoning provisions are modified so that every child has the right to attend the nearest school, and if schools are entitled to enrol any other student who can be accommodated (subject only to an independent ballot where there are too many students wanting to enrol). We have proposed more assistance to parents who want to educate their children themselves. As well, we have set out the circumstances under which parents would be helped to set up a state-funded alternative institution.[99]

As far as the strict business of administration was concerned, the Picot Report had somewhat less direct import for secondary schools than it did for primary schools. In 1987 there were 212 state secondary schools run by 176 boards of governors, 20 private schools (mainly Catholic), and 49 integrated schools. In theory at least, boards of governors already held considerable powers and responsibilities. These included the hiring, firing, and disciplining of teachers, including the principal. The board also controlled the buildings although maintenance and capital works were the responsibility of the appropriate regional office of the Department of Education.[100] Private secondary schools, by contrast, were independently owned and operated, although they could receive a state subsidy for staffing amounting to 25 percent of the salaries paid to an equivalent state secondary school. In addition they possessed administrative independence including the freedom to select and pay staff. Like state secondary schools, integrated secondary schools had boards of governors and teachers' salaries and conditions were identical, though the school's proprietor owned the school and determined capital works for the institution.[101]

It was conceded in the Report that some elements of centralized control were unavoidable for secondary schools, given their pivotal role. It was pointed out that although administered by boards of governors, funding for secondary schools derived from the state and was pre-allocated along set budget lines that were ultimately subject to government approval. PPTA and a government agency known as the Education Service Committee negotiated teachers' salaries on an annual basis, and individual teachers progressed automatically up a nation-wide salary scale. This did not distinguish according to areas of need, although this possibility had periodically been advocated. Individual secondary schools were seen as possessing some discretion in this regard through being able to allocate a given number of "positions of responsibility (PR positions)," though the Picot Report argued that PRs were awarded according to a departmental formula and that, once distributed, they were difficult to take away.[102]

The published report was now a public document. Inevitably, given the intense educational debate of the previous decade, and the undeniably radical tenor of its recommendations, it was about to enter a period of intense public and political scrutiny.

Chapter 7

A Long Way to Go before We Win the Battle

As we have seen, the New Zealand Department of Education's draft statement for the OECD Reviewing Panel saw education as a highly political activity, with well-organized lobby groups enjoying a close relationship with the media resulting in a constant massaging of public opinion.[1] In many respects, however, attempts to both canvass and to shape community and professional views following the publication of the Picot Report were to be unprecedented in both scale and sophistication, even in such a highly politicized sector as education. In the space of a few months a new standard was to be set regarding marketing strategies expressly designed to "massage" public opinion.

"The Year for Education"

Given existing speculation emanating from the media and within the education sector that the soon to be released Picot Report would indeed present radical recommendations for reform, Lange's briefing to the media in April 1988 clearly sought to allay growing concern. The Prime Minister conceded that there were "some quite extraordinary things being said about Picot, as some sort of privatization or saving money on the education vote generally."[2] He emphatically denied that education was being seen as a market commodity, however, because in the government's view there was "no way at all in which you can be equitable and leave it to the market to determine educational access." Turning to the concern that the

Treasury Briefing Paper in particular was raising throughout the education sector, Lange argued that Labour's position did not mirror that of Treasury because the government had a philosophical commitment to the role of government in education.[3]

What was perhaps unknown to those who listened to these comments at the time was that behind them lay a carefully crafted and comprehensive marketing strategy. Evidence from the recently available Lange papers suggests that the need for such a strategy had been foreseen as early as February 1988, before the Picot Committee had delivered its final report. Following a meeting with Ross Vintiner, Lange's press secretary Harvey McQueen, Executive Assistant (educational communication) in the Prime Minister's Office, prepared a brief two-page draft paper outlining what he termed "a proactive strategy to be planned and actioned prior to the report's release that might not only ensure a smoother implementation of the Report's recommendations but also generate a sense of action and achievement in education."[4] Noting that "teacher morale is low and spiraling downward, especially in secondary where they are trapped in a nineteenth-century model that has little relevance to today's conditions," McQueen sought to counter the union strategy of focusing teacher anger on the government by providing a sense of purpose.[5]

> At first there was general pleasure amongst teachers and educationalists that the Government had stopped short-term tinkering until the Picot Report was presented. There is now, however, a growing apprehension fuelled by speculation and scare-mongering about the effects of the changes. If this feeling gathers momentum it could jeopardise the whole reform, create great bitterness, and affect the government's image.[6]

He felt that the campaign to market Picot would have to be initiated immediately, beginning with a media build-up about existing educational achievements. The campaign itself was to give people a sense that they would gain by any changes the report might recommend. It was to be two-pronged, aimed at both parents and public, emphasizing that they would have greater accountability and more responsibility while ensuring them that their problems would be understood. Such a campaign, McQueen thought, might be called something like "THE YEAR FOR EDUCATION."[7] Furthermore,

> to engender a sense of purpose, direction and challenge the campaign will need to stress attainable objectives. Education has been claiming more than it can deliver for decades. If the sights can be lowered this will alter the whole perspective. Teachers would be relieved, they could be human again, no longer attempting the impossible. The public will see set objectives being achieved.[8]

In a slightly later memo to John Henderson, McQueen conceded that the educational constituency would be hard to please, thus it would be difficult to find the right tone. At one point in the memo he appears to have envisaged, "something along the lines of Roosevelt's fireside chats—lets work together to publicize the good things and reform what's not working."[9] However, while McQueen believed that the appointment of a new press secretary (education) and executive assistant (education communication) would have a crucial role to play in any strategy to market Picot, he was also convinced that there was a need for to hire a good public relations firm to undertake qualitative analysis and to prepare a campaign.[10]

This suggestion found almost immediate effect, with the overall responsibility for the Picot marketing strategy adopted by the government being contracted out to Logos, a prominent Wellington Public Relations Firm. The brief for Logos involved marketing the advantages offered by the Picot recommendations to the electorate as a whole. The hiring of private firms as public relations consultants was not, in fact, entirely new to New Zealand education. Following the release of the controversial Johnson Report in 1978 Merv Wellington, the new Minister of Education in Muldoon's National government, announced his decision to hire Link Consultants Ltd to provide an external analysis of the submissions to the report, "because of the obvious independence offered by those outside the education field."[11]

What *was* new on this particular occasion, however, was the extent to which Logos was to be involved in providing direct advice to the government through the Prime Ministers Office. The Lange papers furnish a fascinating insight into the mechanics of what was a carefully crafted, well-run campaign. Its objectives were to target specific groups whose reactions might be crucial to the outcome of the reforms and to influence public opinion in general. Logos' brief was extremely comprehensive, embracing general strategy, client liaison, media monitoring, submission analysis, and briefing of officials as well as information gathering and dissemination. By early 1988, Logos was advised that the government had confirmed its appointment. The firm was to prepare a communications strategy conveying the government's goals in education and assisting in the release of the Picot Report and the collation of the various responses to its recommendations. This was to involve a campaign of some six months duration, commencing on April 19, 1988.

A July 1988 memo set out the objectives of the Logos-managed campaign in some detail. The campaign was

1. To explain to public, and particularly the education sector, the necessity for educational reform and hence the reasons for the Picot Report.

2. To develop a public relations strategy to assist the Education Department in implementing report.
3. To ensure public debate was well-informed.[12]

To this end Logos was to be responsible directly to Lange as Minister of Education. The Department of Education was to be represented only by a small steering committee based within the Prime Minister's Office convened by McQueen. The department was, however, to bear all the costs of campaign from its own budget, with up to $400, 000 being already approved by the steering committee.[13] In the event, the Picot Implementation Phase Budgets and General Strategy and Submission Analysis Fees for June and July 1988 only were to total $381, 984.00.[14]

A brief press statement announced the release of the Picot Taskforce Report by the Prime Minister on May 10, 1988. In his statement, Lange described the report as "a bold and innovative document." He reiterated that he could not, "stress enough the need for us to have an effective and efficient education system that will take New Zealand successfully into the 21st century." Aware of the potential for controversy, Lange also emphasized that it was not a criticism of people working within the profession now, but rather of the system they worked within, adding that, "many of you, I know, feel ham-strung by a system creaking under the weight of decades of ad hoc alteration."[15]

The major national dailies provided full coverage of the release with most of the editorial reaction being favorable. The *Evening Post* observed that the release of the Picot Report completed what it called "the holy trinity": the first two of which had been the Gibbs Report into the health system, and the Royal Commission on Social Policy. The Picot Report was described as having met its brief in finding the Department of Education's administrative layers as having too much power and schools, too little. The editorial warned, however, that there were "some dangers in far greater control being vested at the local level." These included principals needing "the judgment of Solomon" in taking on a director's role, and the capacity of parents to play a major role in the running of schools.[16]

A *Star* editorial expressed a belief that the claim of Business Round Table Chairman, Sir Ronald Trotter, that there had been a 50 percent increase in educational expenditure merely echoed the general public dissatisfaction with schools, which had also been endorsed by teachers themselves. The paper conceded that

> Education Boards, which will get the chop, will welcome the Picot report with the same enthusiasm as a turkey looks forward to Christmas. But

nearly everyone else appears to approve of the almost complete reform of the education system.[17]

In similar vein, an editorial in *The Press* expressed satisfaction that "the existing unwieldy monolith of education administration is demolished." It went on to claim that the release of the Picot Report had

> attracted general approval since its release this week. The Post Primary Teachers Association has reserved judgement on some aspects, a few authorities have doubts about the practicalities of parts of the package, such as the relaxation of the zoning system; and the education boards whose abolition is central to the report do not yet seem to have come out of shock. By and large though, those who have ventured opinions on the report have welcomed the features they personally support and focused their comments on these. The impression given is one of universal satisfaction.[18]

Likewise, the *Northern Advocate* declared that "after the embarrassment of the Gibbs report on health services, because it put economic progress before social issues, and the Royal Commission on Social Policy, for the opposite reasons, the Picot report on education must be a welcome relief for the Government."[19]

Behind the scenes, however, the advice Logos was providing to the Prime Minister's office was rather less sanguine about public reaction. A confidential letter from the firm's Managing Director, to McQueen dated 8 June regarding the proposal to use TVNZ's *Frontline* current affairs program as the initial vehicle through which the government responded to the Picot Report, underlines the strong influence Logos was to have on the way the campaign immediately following the report's release was run, and particularly on the nature of government responses to the various questions that would inevitably be raised.[20] Though Logos reluctantly accepted that decision to utilize *Frontline* in this way was now final, it nevertheless outlined three main areas of concern. First, any release of policy would "be taking place in an environment which will almost certainly include dissenting views." Second, there was a clear danger that the Picot Report would "take on a much greater party political 'campaign' character, possibly destroying some of the goodwill which resulted from the release." Third, Logos warned that the staff whose positions were directly effected would now hear about their future via the news media. The problem here was that, "given that good, direct communications with staff are an important part of our strategy, there is potential to endanger a more hostile reaction."[21]

Logos was understandably concerned to keep this particular aspect of the strategy out the public gaze. Documents deemed to contain commercially sensitive information about how its business operated were thus not to be subject to release under the Official Information Act.[22] Twenty years on, however, this document and others like it offer a revealing glimpse into how the government's largely successful campaign to sell Picot to the wider electorate was actually managed.

In one such document, dated June 1988, Logos specifically identified several target audiences upon which the campaign should logically focus. Teachers were regarded as having the ability to influence local community responses. It was conceded that there were professional issues at stake for teachers, whose response as a group was also seen as having implications for future industrial relations. Department of Education staff were identified as being the most directly affected regarding future employment. This group was also perceived as being crucial for the successful implementation of any changes implemented as a result of the Taskforce's recommendations, particularly as their skills would be critically needed during the changeover period. Educational lobby groups and opinion leaders were identified as an important audience, hence there was a need for them to have, "an accurate understanding" of the issues involved. School Committees and secondary school boards of governors were perceived as needing to have confidence that the new system would work. More specifically, they required reassurance about future training in their new roles, ongoing support, and continuing confidence in the transitional arrangements put in place. "The Community" was described as being a direct audience only at the beginning and very end of implementation phase, because, "after that, provided the opinion leaders in each community are positively motivated, the issue can be expected to disappear below the surface."[23]

Specific instructions on dealing with the media were also issued in this June 1988 document. Thus it was recommended that both Logos and the Prime Ministers office should, "identify and answer all issues raised in the consultative process, even if the issues are difficult and the answers not clear-cut." The justification for an "up-front approach" of this nature was "to tackle problems head-on and, as far as possible, preempt the criticisms of vested interest and lobby groups."[24]

In fact, the initial PPTA reaction to its release was decidedly mixed. On the one hand PPTA president, Ruth Chapman, warned a union conference in Hamilton that the teacher unions were now threatened by a "New Right philosophy" in education. In her view, well-publicized proposals to free up educational structures "were just headlines, behind which lay the New Right philosophy which assumed everyone was motivated by self-interest—money, status, power or some other gain." Chapman regarded this philosophy as being "alien to cultures in New Zealand," having gained ascendancy through business and commercial interests.[25]

On the other hand, despite this sharp reaction, the focus of the term two issue of the *PPTA Journal* in early 1988 was squarely on school-community partnership. Robin Fry's editorial soberly observed that the submissions to the 1987 Curriculum Review had given a clear message that communities wanted more involvement with their schools, and that the Picot Report had responded by laying the onus on the community to help administer education.[26] The feature article, by Massey University academics Roy Nash and Wanda Korndoffer, observed that during, the last fifteen years, there had been a constant debate about education and the community. Fuelling this debate was the need for the state to ensure that the education system was able to produce the skills required to sustain its economic goals, the desire of employers to recruit highly qualified applicants, and the attempt of parents to ensure that their children gained a comparative advantage in credentialing for the labor market. Nash and Korndoffer argued that in this context, community education as a vision became increasingly attractive and, now that the Picot recommendations had been implemented, community control of education would become a fact. They conceded that the whole concept of community education would have to be reconstructed and contested at the local level, with schools having to appreciate coexisting and often competing ethnic and class divisions. Nash and Korndoffer went on to predict that "if the teaching profession sets its face against the thrust of the popular community demands, which Picot does reflect, with a misplaced determination to maintain the *status quo* then it will surely gain nothing."[27] At the same time the authors perceptively recognized that the post-Picot situation could have been much worse. For instance:

> We are to be spared, it seems, an ideologically and politically motivated deregulation of the education system of the sort advocated by Treasury, a policy which would have allowed only the maximization of private individual interests and done nothing to serve the public interest of social equity in education, and are instead to gain a more community centred system of educational administration. This is certainly to the good for any attempt to expand private and quasi-private education, for example on British or Australian patterns, ought to be resisted. Yet, as Picot has recognised, if communities are to be empowered that requires, among other things, that they should gain some definite communal control over the institutions which supply education and training.[28]

Moreover, although Lange had often expressed his confidence in the Picot Committee, he had also reiterated that that the government was not necessarily bound to accept all its recommendations, and that the definitive policy document was yet to come. In fact, the three months from the release of Picot Report in June and the release of *Tomorrows' Schools* in

August was a period of intense activity and discussion in which the media became a vehicle for assertion and counter-assertion, much of it being used to either reassure the public that the government was not committed to the extremes of radical reform or to influence key decisions on the report's implementation. Dr. Lockwood Smith, National's spokesperson for education hinted at a bipartisan approach on education if the government fully adopted the Picot Report, leading to an emphatic denial from Lange on National Radio.[29] Moreover, visiting primary and secondary schools in Christchurch soon after the report's release, Lange clearly signaled his intention to back away from some of the more controversial aspects of the report, even conceding that in some respects it may have gone "a bit far."[30] Of particular concern to the government was the Picot Report's observation that schools might well exercise discretion in spending devolved funds. The prime minister's assertion that schools would not be permitted to employ fewer teachers and purchase extra computers instead, however, appeared to expressly rule out total bulk funding.[31]

Recognizing that the Picot Report would inevitably be controversial, Lange had as early as April, proposed the establishment of an officials committee consisting of himself as chair with representatives from Education, State Services and Treasury, to advise Ministers on the reform of education once the Picot Committee had presented its report. The officials committee would also coordinate responses from departments, oversee the communications strategy for the release of the Picot Report once it had been completed, and assist in the preparation of a government statement on educational administration.[32]

In the event more than 20,000 public responses to the newly released Picot Report were eventually received. Given the scale of the response, weekly reports were prepared for officials to monitor public reaction to the various issues raised. A final 22 page report dated July 13, 1988, observed that two samples had been taken of responses, each of approximately 1,000 returns. The samples classified responses as coming from teachers (26/25%); Department of Education, education boards and their staffs (18/19%); specialist and advisory teachers (11/10%); school committees and boards of governors (14/12%); users of specialist services (14/12%); Māori (6/7%); integrated schools (1/1.5%); and unspecified (3/2.5%).[33]

Perhaps surprisingly, the relationship of the individual learning institutions both to the state and to the other proposed agencies appears to have attracted little public comment, perhaps indicating that the basic structures and philosophy proposed by Picot were acceptable to many respondents. BoTs, however, emerged as a dominant issue. Although there was general acceptance of the concept, there were specific concerns raised about the extended role of parents in educational administration, including the time BoT members would have to spend on administration, the skill level required, and their degrees of accountability/liability. Other questions raised involved the disparity in expertise and resources schools could draw upon, and the dangers of boards either being un-representative of their communities, or being "hi-jacked" by determined minority groups.[34] Teachers in particular, were concerned about the power and autonomy of BoTs, and remained determined to retain national conditions of employment in the face of the threat they perceived from any implementation of bulk funding.

In addition to the submissions, many members of the public chose to write letters directly to Prime Minister Lange. A Logos report claimed that by mid-June no fewer than 2033 letters had been received. The largest single group of letter responses was from teachers, followed closely by principals, school boards, and committees. Although generally reflecting the tenor of the more formal submissions, there was a strong representation from rural communities (approximately one-third of the total). Although there was general support for the Picot concept of devolution, major concerns were raised about the availability and funding of specialist education and advisory services, as well as doubts about the ability of rural areas to attract the necessary level of professional and administrative expertise to sustain BoTs.[35]

Within education, sector responses were divided. Although the primary teachers union, the NZEI remained implacably opposed to the report's implementation, the PPTA accepted the need for reform.[36] While long experience with local control in the form of boards of governors and their equivalents probably softened PPTA's overall reaction at this juncture, however, the union was nevertheless concerned that the proposed devolution had gone too far. Its submission pointed out that greater choice did not necessarily lead to greater equity for students, asserting that state guidelines would be needed to guarantee equity principles were upheld. It also questioned the availability, willingness and competency of potential BoT members, disagreed with the separation of policy and operations within the proposed new Ministry of Education, wanted teaching salaries and conditions of service to remain nationally determined, and criticized the lack of emphasis on curriculum matters in the report.[37]

The reaction from other secondary school-orientated organizations was one of general support for the Picot proposals, although questions were raised over their details. The Secondary Principals' Association recommended that BoT members should be current parents and expressed some concern about principal/board relationships, but also queried why teachers should not be placed on contracts, as had been proposed for principals. The Secondary School Boards Association on the other hand, strongly supported the objectives and core values of Picot. The Association approved of the recommendation that the majority of BoT members should be parents, but felt that existing contradictions in the report regarding the relative power between boards and teaching staff should be resolved in favor of boards. The Independent Schools Association likewise generally approved of the Picot recommendations, particularly the key concepts of parental involvement, local decision-making, parental choice, and accountability. The Association did, however, express a strong desire for further negotiation over the issue of how far private schools might be obliged to follow national guidelines. The Association of Proprietors of Integrated Schools also supported the general tenor of the report, while emphasizing the necessity of retaining the general principles of the Integration Act. The New Zealand Principals' Federation saw many of the report's recommendations as increasing the autonomy of principals, but expressed some misgivings over the proposed appointment of principals, and the danger of board capture. The New Zealand Area Schools Association, representing rural primary and secondary interests, endorsed the report's general direction, but nevertheless expressed some concerns over equity for rural schools.[38]

Three major concerns about the Picot recommendations were noted by the Officials Committee. Perhaps the most pressing of these was the position of Māori education in the post-Picot era. On July 1, 1988 Wiremu Kaa of the Department of Education had tabled a combined submission from Māori, collated from special team of advisers, inspectors, and education officers presently working in Māori Education. The team had held meetings across the country from Te Hapua in North, to Invercargill in the South, involving 222 hui with 7,189 attendees.[39] The submission was thus able to assert with some confidence that

> with one or two exceptions, the Iwi support the spirit and intent of the report and the numerous views expressed were largely positive. A small group rejected the report in total. All Iwi, however, were concerned that the status of the Tangata Whenua, the Treaty of Waitangi and a Māori presence was absent in the Picot structure. A common question raised was "kei whea tatou nga Māori" in the Picot structure?[40]

The combined submission asserted that Iwi wanted Māori education to be managed by Māori people, but conceded that this desire had been expressed in different ways. These included establishing a secretariat of Māori education within the Picot structure to develop, monitor, and evaluate Māori education, ensuring Māori representation at all decision-making levels, and creating an entirely autonomous system independent of Picot.[41]

A number of issues and recommendations incorporated in the combined submission related to the proposed new Boards of Trustees (BoTs). Many Māori had expressed the fear that "white collar people" might take over and dominate BoTs. Continuing language loss and the gap between Māori and non-Māori student achievement remained major concerns. Thus, although there was an interest in the Picot concept of "schools-within-schools" to cater for minority interests, there was also a worry that this solution might lead to loss of control, and a perpetuation of the achievement gap. Turning to Māori educational initiatives, the combined submission argued that the Picot Report was "weak on biculturalism." In its view Kaupapa Māori schools and bilingual schools should have been entitled to the same support as integrated schools in regards to special character. Te Kohanga Reo and post Kohanga children had to be better provided for, with Māori culture and development being specified in policy and determined by the Māori community. There was also the feeling that schools with predominantly Māori mokopuna should receive similar guarantees.[42]

The second question confronting the Officials Committee involved the difficult question of special character and its legal status in the post-Picot era, especially in relation to Catholic Schools. Writing in response to a query from Ross on what changes would need to be made to the Private Schools Conditional Integration Act 1975 in light of Picot, G. Knight noted that the Picot Report made brief passing references to integrated schools, apparently envisaging a continuation of integrated schools as district institutions. Knight warned, however, that the Report did not have a clear statement as to how integrated schools were to fit into new system.[43] Moreover, the Report did not give integrated schools the special protection and status of Integration Act, such as a guarantee that special character would be preserved or protected.

The third issue facing the Officials Committee involved the election and composition of BoTs in the new structure. As the most fundamental change recommended by the Picot Taskforce, questions concerning the implications and operation of BoTs had been raised by a high proportion of submissions and letters. These centered around the application of Treaty of Waitangi principles, the representation of women and minority groups, the safeguarding of teachers' salaries and conditions, the exact nature of the relationship between boards, principals and teaching staff, liability concerns, matters surrounding bulk funding, the difficulties of attracting suitably qualified members in rural areas, and student membership of BoTs.

The revised report of the Officials Committee, released on 24 July and sent on to the Cabinet Social Equity Committee, reflected these concerns. It recommended the adoption of the Picot Report's general principles, including its criticisms of the current bureaucratic structure of education, and the creation of a new Ministry that would implement educational policy but not provide educational services. Bulk funding was supported in principle, but with equity safeguards. However, officials wanted measures to be put in place to "ensure a good representation of women, Māori, Pacific Island and working class parents on BoTs" and that the present "middle-class capture" does not continue."[44] They also felt that alternative selection processes for Māori and Pacific Island parents must be considered but as an interim measure, co-option might be necessary to ensure better representation. Equity issues were to be integrated into all aspects of the reforms, and the new system was to ensure greater equity for women, Māori, and other disadvantaged groups.[45]

Officials were aware that the term "special character" could take on a wider meaning as parents took the opportunity to establish schools, as would be permissible if the Picot recommendations were accepted. They were, therefore, concerned that the definition of "special character" be extended to include kaupapa Māori schools, bilingual schools and other schools where, in the view of the Minister of Education, such a special character existed., Single sex schools, however, were recommended to remain as a separate category of general schools.[46]

Finally the Officials noted the considerable sums of money to be handled by BoTs. They opined that legally speaking, members would currently be liable for "breaches of trust," as was the case with trust businesses. Accordingly, they recommended that such a high standard not be applied to BoTs given both the need to avoid discouraging wider membership, and the presence of national guidelines and the Review and Audit Agency (RAA).[47]

Meanwhile, with public and professional opinion apparently hanging in the balance, a further aspect of agreed marketing strategy between Logos and the Prime Minister's Office was to commission an extensive campaign of public pulse-taking. Insight New Zealand, an Auckland firm, was subcontracted to run a telephone survey ascertaining public opinion concerning the major Picot Report recommendations.[48] Accordingly, Insight carried out a telephone survey of 750 people between July 18 and 27, 1988, allocating approximately ten minutes for each interview. The responses, passed on to the Prime Minister's Office, were revealing. Some 43 percent claimed to be dissatisfied and 40 percent satisfied, with the current state of education. Most people (69%), claimed to have heard about the Picot Report either through radio, television or newspapers. Regarding the key recommendation that BoTs consisting mostly of elected parents would be responsible for overseeing the operation of the school, 47 percent approved and 43 percent disapproved. However, 55 percent of those surveyed accepted the proposition that BoTs would control the school budget and determine how resources were allocated, with only 37 percent disapproving. An overwhelming majority (79%), were unhappy with the notion that the core subjects that students studied, such as English and Mathematics, would be compulsory at the national level, while some other subjects might be set locally. Likewise, a considerable majority (70%) rejected the view that BoTs would be able to determine how much teachers were paid. Proposals for a Parents' Advocacy Council and Community Education Forums met with strong approval from the sample.[49]

In late July 1988, Insight New Zealand supplemented this information with a comprehensive summary of the key findings and recommendations arising from the National Survey.[50] It was conceded that

> for the Picot Report and its recommendations to be accepted by community, a major communications campaign, targeted at the general public, will be required. The contents of Picot and its suggestions have yet to substantially register with most members of the public. While many people are aware of the report (69%), and know something about it, they have not yet absorbed sufficient information to form even the most rudimentary position. A massive 50% of those who have heard of Picot are presently undecided about it.[51]

The extent of detail in this summary reveals the extent to which the battle for the hearts and minds of New Zealanders was regarded as being crucially important for the future of the reforms. That this battle was far from being over was clearly spelled out. Although Insight believed that the battle for the elite audiences, teachers, teaching professionals, had already been won,

> the overall figures are disturbing, 27% approve, 24% disapprove and 50% are still undecided. The results indicate we are on the right track but have a long way to go, before we win the battle. Support is strongest among white collar workers, upper income earners and those whose children are at school.[52]

However, it was Insight's view that

> most of the 50% of undecideds will be relatively easily reached and convinced. While still to endorse Picot they are generally sympathetic to most of the Picot positions and have no real objections to what he says. Most hard core opposition has already dropped out into the 24% who are currently opposed.[53]

Insight's conclusion outlined the dangers that still lay ahead in winning the battle for minds.

> Our position is reasonable, but we are vulnerable. A lot depends on the opponents of Picot. If they do nothing we will probably be okay. A strongly directed clearly focused campaign in opposition to Picot, could knock us for six, if we did nothing.[54]

In fact, a number of Left-liberal publications were already attempting to marshal public concern. Writing in the feminist magazine, *Broadsheet*, Auckland secondary school teacher Jill Brame epitomized Left-liberal reaction to the Picot Report, while also underlining its essential contradictions. Brame emphasized that Brian Picot was a Foodtown supermarket director. She believed that the structure set up by Picot at first glance suggested that the existing "Pyramid of Power" had been turned into what she termed, "the Plumber's Plunger of Power," which on closer analysis better resembled a bottleneck where a small oligarchy now exercised power over many. Issues such as biculturalism, equity, and resources were left unresolved. As for devolution,

> there is a charming and idealistic assumption behind the report that people in general are rational, objective, idealistic and committed educators with a wide variety of communication and management skills. It is rather pleasant and naïve and gives me the warm fuzzies for all of two minutes.[55]

Citing a mother of five she had interviewed, Brame asked why, if people paid taxes for students to gain an equitable education, were parents being hired to do the job of qualified people, a number of who were now facing unemployment. In a probably unintentional ironic reversal of Left-liberal and neoliberal critiques of defense, foreign affairs, the police and hospitals throughout the decade, Brame now emphasized the case *for* autonomous professional expertise, justifying faith in education professionals as the best people to run education in the same way that armies should be "run by people who know how to run armies."[56]

Business criticism of the Picot Report following its release also sought to influence public opinion. In so doing, they displayed a similar ambivalence to its call for local control of schools by community representatives. Critiques emanating from this quarter also attempted to draw a distinction between the intentions of the Picot Committee, and the subsequent capture of the process by education bureaucrats, supported by teachers' unions, who sought to retain a large measure of centralization. Writing in the *National Business Review* Allan Levett, a Wellington consultant in development finance, praised the Picot Report for emphasizing that every learner should gain maximum individual and social benefit from the money spent on education, and that education should be fair and just for every learner. He was critical, however, of the failure to adopt clearer national objectives for education, including the neglect of "the economic survival aspects of society's requirements." Levett was particularly concerned that the Policy Planning Council would become "just another quango that would ultimately fail to adequately to coordinate levels of attainment in various subject areas,' especially mathematics, science, Asian languages and a range of design and communication skills that would tie education directly to the country's economic and trade requirements."[57]

This desire to alert the public to the possibility of the Picot Report being subverted or taken over at the eleventh hour was not just confined to Left-liberal educators or Right-of-Center business interests. Lange, having recently returned to work following hospitalization for heart trouble, was soon publicly claiming that his previous warnings of drastic spending cuts and higher taxation in the immediate future had been made partly for political reasons. In a scarcely veiled reference to what he regarded as backroom dealing on the part of Finance Minister, Roger Douglas, Treasury and SSC, Lange claimed that he had hoped "to get my own back on the 'New Right'" because they had gone too far in attempting to promote public policy.[58] Likewise, a concern to alert the education sector to the possibility that right-wing views might yet find expression in the definitive government response to the Picot Report appears to have been a strong motivation behind the subsequent public disclosures from Picot Taskforce member, Professor Peter Ramsay. In late July Ramsay went to Wellington at short notice to discuss aspects of the report with members of the Officials Committee. In a prerecorded address to a Hamilton education seminar at the University of Waikato, Ramsay claimed that officials from Treasury and SSC had "made strenuous attempts to influence the Picot review" Both had stepped outside their portfolios and now "had their tentacles well and truly entrenched" in educational decision-making. Treasury and SSC had come with "well-prepared plans" that they had placed before the Committee at an early stage, and the Taskforce's own views had emerged only subsequently. Ramsay emphasized that he was raising these issues now because a number of aspects in the Picot Report would be opposed by state officials. Hence, "it would be an interesting exercise in the next few months to see whether the Government takes on board the proposals offered by the task force or whether the Treasury and SSC view becomes the predominant one."[59]

Thus it was that *Tomorrows' Schools* was officially released in early August 1988, amidst a storm of controversy. Lange and the government clearly regarded this document as being the definitive blueprint for education reform that presented the policy decisions arising from the earlier Picot Report. Hence, in launching it, Lange emphasized that it was "in line with Govt's [sic] overall objectives of having decisions made as close as possible to the actual point of implementation."

> Ordinary people have increasingly felt powerless before the professionals in health and education and elsewhere.... In the past, involvement has often meant having a say. But the professionals still made the decisions. This system will give the parents more than a greater say. Not just parents. Governments in different ways as well. Until I became Minister of Education I had little idea how circumscribed this office was. I cannot even visit a secondary school without permission. So the mechanisms proposed by Picot and with quite a few adaptations or modifications accepted by the government will spread the authority in education. It's now a three-way partnership, parent, educator and Government. And that's how it should be.[60]

In his foreword, however, Lange asserted that although much of the document was "an affirmation of the Picot proposals... in some areas the Government has chosen to depart from the taskforce's recommendations."[61] The Report of the Officials Committee and the concerns of Cabinet's own Equity Committee were clearly reflected in the assertion that equity issues in general and Māori interests in particular were to be more carefully addressed than in the Picot Report. Equity objectives were now to underpin all policy related to the reforms. The new system was to progressively achieve greater equity for women, Māori, Pacific island and other minorities. It was to ensure that equity issues were integrated into all the changes rather than treated as "an optional extra." The features of the existing system that had promoted equity were not to be lost, and in the new system there was to be a monitoring of progress toward equity goals.[62] The heightened emphasis on equity did not please everyone. Picot, for instance, was later to claim that this stronger emphasis on equity meant that school charters turned out very differently from what the Taskforce had originally envisaged, overemphasizing equity and underemphasizing excellence.[63] Moreover, the tensions between these two goals were to abruptly resurface when National came to power following the 1990 general election.

In addition, the composition of BoTs was to be mandated to better reflect equity principles and avoid the danger of "capture." Each BoT was to consist of five members elected by parents of students at the school, the principal, a staff member to be elected by the school staff, and a member elected by the student body in the case of secondary schools. Integrated schools were to have two further members nominated by the proprietor. In addition to their elected members, BoTs were given the authority to co-opt a maximum of four further members, "to ensure that the board properly reflects the composition of its community, or to ensure that particular expertise is represented and available to the board," although none of these was to be a teacher employed by the board. BoTs were also to "encourage a fair representation from their community—especially from women, Māori, Pacific island and working class parents in communities where this is appropriate."[64]

Tomorrow's Schools fell short of imposing bulk funding as such. All funding was to come to schools as a bulk grant, but this was to have two distinct components—teaching salaries and operational activities-each based on a separate funding formula. These were to be sensitive to local needs and would be weighted in favor of both equity considerations and rural schools.[65] This compromise brought criticism from both critics and supporters of bulk funding. The teachers' unions saw it as the thin end of a very considerable wedge that would lead to the erosion of both teaching and learning conditions. Treasury viewed the failure to introduce bulk funding as nothing less than an undermining of the independence of schools.[66] Picot Taskforce member, Whetu Wereta, considered it "a pity" that bulk funding had not been adopted.[67]

Māori interests, however, were to be assured through several further measures. There were to be opportunities for children to be educated in the Māori language if parents desired it. As with other parents, Māori parents would either be able to homeschool their children or establish their own institutions where the system was not meeting their needs. Individuals within whanau were eligible for election to BoTs and the close partnership between BoTs and locals communities could be further strengthened during charter negotiations.[68]

Tomorrows Schools also placed an increased emphasis on developing national guidelines for education. These guidelines were to be "the means of setting, maintaining and developing national standards of achievement in education, and (would) be an expression of matters of national interest."[69] The guidelines were also to include an expression of equity principles, national curriculum objectives, together with codes of conduct for BoTs and principals.

These and other modifications outlined in *Tomorrows' Schools*, however, went too far for some critics, and not nearly far enough for others. Some within the National Party were clearly unhappy with the increased emphasis on equity issues which they regarded as being at the expense of standards. When National eventually succeeded Labour on the Treasury benches, they therefore sought to tip the scales in favor of competition and excellence.[70] Some Māori on the other hand, believed that Māori opinion had been deliberately manipulated in the interests of sustaining what remained an essentially Pakeha system of education.[71]

Further opposition came from some commentators from within the education sector. With the release of the definitive government response to the Picot Report, the specter of a clearly focused campaign in opposition to Picot predicted by Logos now showed every sign of materializing. A hardening of secondary teacher attitudes was more clearly evident in the Term Three issue of the *PPTA Journal*. In the wake of the publication of *Tomorrow's Schools* in August 1988 and the realization that the Picot Report's recommendations had been largely accepted by the government, this offered a much less conciliatory tone.

A much more critical Fry editorial warned readers that to innovate was not to reform. Recent British and Australian educational reforms were viewed as having threatened teachers' rights, with the New South Wales education minister's "draconian" reform proposals having "united students, parents and teachers in Australia's biggest demonstration since the Vietnam war protest."[72] John Codd, then a Reader in the Massey University Education Department, deemed the sense of urgency that had accompanied the production of the Picot Report "alarming." Although Codd conceded that the influence Treasury had in the preparation and writing of the Report was a matter for further research, he noted that a Treasury officer had been seconded to the committee as a part-time secretary, that the Treasury had released its own had released its own document on education before the Taskforce began writing its own report, and that the Picot Report and the Treasury document were similar in emphasizing that a key element in addressing equity and efficiency was empowerment through choice and information, for families and individuals as education customers. In short, Codd viewed devolution as little more than a cynical attempt to concentrate the real power of decision making at the center, while deflecting public dissatisfaction away from government, where it would be "vented" at the local level.[73]

In similar vein Colin Lankshear, then a lecturer in the politics of education at the University of Auckland, argued that, "to understand the Picot Report we have to understand the logic of privatization as a recipe for curing current economic problems, and to see how this logic links up with New Right thinking about education." A specialist in literacy politics and an active supporter of the Nicaraguan revolution Lankshear opined that the mentality of the market that had led to Guatemalan children being sold as organ donors in a global context where third world children were viewed as sources of essential raw materials might well see the Picot Report aftermath producing its own brand of market-led horror leading to a unique form of human deprivation, degradation and decline.[74]

This last view in particular left little or no room for negotiation or for compromise. However, although the majority of the commentators to this PPTA volume now emphatically rejected the whole concept of reform embraced by Picot and *Tomorrow's Schools*, several still revealed a degree of ambivalence concerning the new structures that had been noticeable during initial reaction to the document. Graham Hingangaroa Smith for instance, affirmed the recognition the Picot Report gave to the need for fundamental educational change, but argued that it had failed to deliver meaningful change to Māori. The Report's apparent emphasis on devolution and community decision-making were viewed as having ignored Māori cultural and language aspirations while failing to recognize the special position of Māori as tangata whenua. Instead,

> every school becomes a site of contestation; every school has license to hold a referendum on what support is given to Māori interests. The goodwill and support that already exists within the teaching force for Māori needs and aspirations will be tested under the new Picot structures, in that an increased dependency is placed on these mainly Pakeha teachers and principals to show even more goodwill and support to maintain the position as it stands now.[75]

This view neatly encapsulated both the concern and the dilemma both Left-liberal and neoliberals were to articulate in reacting to the devolutionary tenor of the Picot Report. Once again, we see here a central dilemma of devolution in striking an acceptable balance between setting national priorities (however these are defined), and accommodating local variations. The recognition of this dilemma was clearly evident in an article by PPTA advisory officer Phil Capper. Capper questioned whether an administrative system based on teacher accountability and a balance between central government and local institutions would bring the country any closer to what he and his fellow contributors would have regarded as a fair and just system for every learner.[76]

> If I say that the mechanisms for establishing teacher accountability that are proposed will not work I am declared to be against teachers being accountable. If I say that the proposals for empowering Boards of Trustees are fraught with difficulty I am declared to be opposed to community participation in educational management.[77]

A similar ambivalence ran through the articles in this volume of the *PPTA Journal* by Ann Dalziell regarding rural schools, and by PPTA Women's Officer Helen Watson, concerning women and girls.[78] Dalziell conceded that *Tomorrows' Schools* "seems to espouse many of the things PPTA believes in," but that devolution and local control could create considerable resourcing problems for rural areas leading to serious inequities.[79]

Likewise Watson found *Tomorrow's Schools* to be somewhat contradictory for women and girls because although on one hand, "there had never been such an official commitment to equity principles," she believed that the demands of women for equal opportunity required, "strong central intervention and support while the philosophic thrust of the changes are away from central control."[80]

This view, however, appeared to run counter to what virtually every educational critic had been advocating with increasing force during the previous decade or more. And as Nash perceived at the time, it was both tragic and unwise for those within education to support centralism against individual freedom of choice, because, "if we react defensively, as statists and upholders of the status quo, then I believe we will lose the popular debate—and deservedly."[81]

Chapter 8

Only Half a Policy

Creation and Continuity

The strength of professional opposition expressed through the teacher unions following the release of *Tomorrows' Schools* inevitably brought casualties in its wake. The prime minister's own sister was one of a number of teachers who felt compelled to resign her membership complaining, "that the PPTA has suggested we all spend time today whilst on strike writing letters to all sorts of people, including the Minister of Education."[1]

Meanwhile, despite the union campaign, the creation of a new Ministry of Education (MoE) to replace the former Department proceeded with considerable haste. A memo promulgated in early November 1988 emphasized that much had already been indicated in the government's decisions on *Tomorrows Schools* and in State Sector Act that provided a basis for determining MoE structures.[2] Hence, an extremely tight timeframe called for an agreement on broad divisional structures by November 21, the preparation of a mission and functions statement by November 28, and a decision on organizational values philosophy and style by December 5. Advertisements for the ministry's first CEO were to be placed by February 15, 1989, with the appointment of both the CEO and the General Manager of divisions by May 15. Not for nothing did an internal memo warn that there was little room for slippage to meet the planned start time of October 1, 1989.

MoE was officially formed through Order-in-Council on May 22, 1989.[3] Senior appointments to the new agency were quick to follow. In early July 1989 the State Services Commission publicly announced the appointment of Dr. Maris O'Rourke as MoE's first CEO. Then coordinator of the former

Department of Education's "Before Five" project, the forty-eight-year-old had previously been Regional Senior Education Officer (early childhood) in the department. Born in Scotland and educated in England, O'Rourke had been a Fulbright scholar, studying at the University of Kansas and at Auckland University, gaining her doctorate in 1986 and going on to lecture at the Auckland College of Education where she initiated an innovatory bicultural and bilingual early childhood course.[4] Although former director general of education, Russ Ballard, was to move out of education to become the new CEO of the Ministry of Agriculture and Fisheries, a number of senior officers in the former Department of Education gained key positions in the new organization. Tony Leverton was appointed general manager corporate services, having previously been executive officer for the Implementation Unit. MoE's chief financial officer was John Gill, the former department's director of finance. Bob Garden became director research and statistics from his old position as director, Research and Statistics in the former department.

A number of key ERO appointments were also ex-departmental officers. Wally Penetito was appointed ERO development manager, having previously been Senior Education Officer (Māori), in the National Regional Office of the Department of Education. Tony Cross became MoE's Auckland regional manager (from being District Senior Inspector [Secondary], in the National Regional Office of the former Department (NRO)]. David Hood became Hamilton regional manager having been education officer, Transition, Hamilton. Val Ferguson was appointed Wellington regional manager (currently senior inspector [primary schools]) seconded to Ministry Establishment Group. Ken Foster was ERO Christchurch regional manager (currently acting director of Schools Division in Department of Education). Mark Canning was MoE's new manager corporate services (currently director of administration, Department of Education). These senior MoE and ERO appointments signaled that there would be a considerable degree of bureaucratic continuity and undoubtedly lent weight to the charges subsequently leveled at the Wellington-based educational bureaucracy by Left-liberal and neoliberal critics alike. The writing of school charters was also well under way by late 1989. Despite the apparent flexibility charters seemed to promise, however, Codd and Gordon have shown how the avenues for a transfer of real power to local communities were progressively tightened, resulting in little real flexibility for schools and the continued maintenance of control by the state.[5] A good indication of how effective this process could be is revealed by the fact that, by September, of well over 400 schools in Hamilton Education Board area, only six were noted as having charters that might require closer scrutiny by the Ministry. As we have seen, however, the tightening of state control over charters stemmed from *Tomorrow's Schools* rather than the Picot Taskforce, largely as a result of concerns expressed by Cabinet's Social Equity Committee.

The Battle Continues

An early indication of the approaching battle over the size and extent of MoE responsibilities and functions was revealed in a memo from Associate Education Minister, Phil Goff, to Cabinet's Social Equity Committee in late 1989. Both Treasury and SSC were clearly concerned with the lack of specificity in MoE's mission, and with its assumption of broad responsibilities. Thus although Treasury had little direct involvement in the initial formation of the new Ministry, it was, along with SSC, to be directly involved in drawn-out controversy over Ministry outputs.

Although Treasury confirmed that the outputs had already been agreed to at the interdepartmental level, it was thought that the mission statement was too broad. Accordingly, Treasury believed that it should "be redrafted to reflect the Ministry's major outputs which comprise the provision of quality, professional policy advice and the efficient carrying out of those operational responsibilities allocated to it by the government."[6] Likewise SSC concurred with the Ministry's mission statement and broad functions, but warned that "as result of Cabinet decisions taken at various stages during the implementation of the education reforms, the Ministry now has more outputs and a more complex structure than was envisaged in *Tomorrow's School's*." SSC also pointed out that the provision of learning materials was originally to have been carried out by a separate Resource Business Unit outside the Ministry, while funding for school transport was to have been supplied direct to BoTs. Moreover, according to SSC, MoE's proposed structure of ten divisions had "the potential for conflicting lines of accountability and might well need to be subject to review in the future."[7]

Pressure was to steadily build during the election year of 1990, with Labour increasingly divided on major policy issues. The 1990 Lough Report, which had representation from both Treasury and SSC, was set up to review the education reform process and to recommend any necessary improvements.[8] Although the report conceded that the transitional period had been tough, it nevertheless argued that "a perception that has emerged in practice is increased central bureaucratic control, increasingly burdensome administrative tasks, inadequate resourcing and support for institutions, and inadequate attention to educational outcomes."[9] Among its key recommendations were the requirements that "the Ministry of Education be restructured to reflect better its core policy role through amalgamation of existing 12 divisions into 3 main groups—policy, operations, Finance and Support";[10] that the Ministry "as part of restructuring group" together those operational activities which do not form part of the Ministry's core policy role "so that they can be spun off in the future"; and that the government "redirect significant proportion of funding from ERO, to schools, reducing ERO personnel to approximately half the present establishment."[11] Not for nothing, did the bitter phrase, "you've been Loughed," come to enter the vocabulary of state servants soon after the report had appeared.

Further pressure for decentralization continued during the lead-up to the election. In September 1990 the Assistant Commissioner of SSC, Paul Carpinter, set out in a twelve-page paper, the state sector principles his organization intended to uphold whatever the outcome of the coming general election.[12] Equality of opportunity, he stressed, did not mean equality of outcome, which in any case was incompatible with both market-based economies and with non-market-based economies such as those of Eastern Europe.[13] In phrases that anticipated the Porter Project, Carpinter observed that New Zealanders traditionally expected a quantity and quality of government service that was up to the standard of, if not better than that available in the most developed countries. The reality, however, was that the country's economic performance was closer to that of Spain than to the United States, Japan, or Germany. It was, therefore, "reasonable to assume that the relative quality of public goods and services (would) reflect this reality."[14] It followed that several "second order principles" should underline SSC's policy advice: minimal intervention, responsiveness, devolution, non-arbitrariness, and similarity. Citing the state's emerging relationship with Māori, Carpinter opined that responsiveness particularly referred to the desirability of a public service that was neither remote from community it served, nor captured by its client groups or providers, while devolution itself involved significant accountability issues.[15]

With the country's first Mixed Member Proportional (MMP) parliament shortly to become a reality, the new state services commissioner, D.K. Hunn, set out SSC's overview of the key issues facing an incoming government.

> The Government faces a threatening economic environment and within the context of a government response to this, the single most important issue is the fiscal deficit. A substantial deficit is certain for the next financial year unless—and probably even if tough decisions are taken by the end of this calendar year. But the immediate problem itself may only be a symptom of

the underlying pressures driving public expenditure upwards, such as an increasing non-wealth generating proportion of the community through high unemployment and an ageing population. Public expectations of what the state sector can be expected to achieve are not realistic in terms of the resources available to it.[16]

SSC's brand of neoliberalism envisaged a core public service comprised of a number of small to medium sized departments (police, defense), supported by a second layer of state sector organizations providing goods and services from organizations legally separate from them and operating as state owned enterprises. Carpinter believed that a major obstacle to achieving these goals were public service chief executives, who might well oppose changes that would reduce the size or influence of their departments. As the main provider of community services, he argued, the state had created a belief that the Crown was the only appropriate provider—a particular perception in education and in health.[17] The pressure for change would, however, become increasingly urgent as the scale of the deficit became apparent and the media began to influence public opinion here. Here, Carpinter noted that the financial markets had already demonstrated their ability to take flight from the New Zealand dollar.[18] The only workable solution lay in establishing a smaller, more efficient and effective state sector, developing strategies in the form of corporate plans, producing accountability systems that monitor and rewarding departmental and individual performance. Hence Carpinter saw a need to recruit and retain appropriate skills, particularly in management, creating a common set of values across all layers of sector which concentrate on meeting needs of identified clients.[19]

National Comes to Power

With critics on both sides of the political divide already publicly condemning the apparent perpetuation of centralized decision-making, further devolutionary pressure was applied to the new structures. This became even more pronounced following the November 1990 election of a National government, led by Jim Bolger, pledged to free schools from what it saw as the excessive constraints imposed on BoTs by the previous Labour government. Although tensions were soon to surface between economic moderates and economic radicals, the new Cabinet quickly agreed on the necessity for substantial spending cuts. National's minister of education, Lockwood Smith, was strongly committed to reducing educational expenditure. In his first formal speech as Minister, Smith told the Association of Polytechnics Conference that he intended to foster "a new level of efficiency in the administration and management of educational institutions," and to "deliver more funding to meet the learning needs of students, rather than to fund central or regional bureaucracies."[20] While other commentators on this period have noted the pressure for fiscal restraints and its impact on education under National, however, the bitter and extended battle this provoked between the MoE on one hand, and Treasury/SSC on the other has been largely ignored.

Pressure on the new educational structures from outside education, particularly from the business sector, soon made itself felt. Released in December 1990, some eight months after the Lough Report, the Sexton Report came to similar but more emphatic conclusions regarding decentralization.[21] Stuart Sexton was the director of the Education Unit of the Institute of Economic Affairs (IEA), in London, and had been a special advisor to two secretaries of state for Education and Science in the Thatcher government. Sexton argued that

> overall, the picture is one of good intentions by Picot for the decentralisation of school's administration, leading to more effective management and more effective delivery of education. A pulling back from this intention by the administrators/politicians to retain a still over-centralised, bureaucratic system followed. One area of policy, however, which Picot inadequately addressed was the creation of genuine consumer choice, which would lead to market pressure to raise educational standards. Decentralisation to the schools is only half a policy; the other half is that those schools must have the means and the *incentive* to respond to consumer choice.[22]

In particular Sexton pointed to continuing forms of central control evidenced in the creation of a new Ministry of Education, a Review and Audit Agency, a Parent Advocacy Council, a New Zealand Qualifications Authority, a Teacher Registration Board. Curriculum and a centralized, prescriptive curriculum Moreover, "perhaps most insidious of all, the charter which each school has to adopt, and which Picot foresaw each school working out for itself, is almost totally written for the school by central government."[23] As we have seen, Left-liberal critics were raising precisely the same issue at this time. The culprit in Sexton's view, however, was not Treasury, but the former educational establishment that had surreptitiously worked to undermine the reforms.[24] Observing that many former departmental officials had got their old jobs back in the new ministry or its agencies, thus perpetuating entrenched bureaucratic attitudes, Sexton alleged that politicians had gained votes by placating the vested interests of teachers and educational administrators, while ignoring parents.[25]

The emerging conflict between MoE and Treasury/SSC centered particularly on the extent of further downsizing; and specifically on the possible privatization of Ministry agencies such as Payroll Services and Learning Media. Battle lines were to further harden when in February 1991: Carpinter outlined what he regarded as some major issues confronting education in a paper intended for discussion by the Education, Science, and Technology Official's Committee.[26] In this paper, Carpinter emphasized that the Expenditure Cabinet Control Committee had noted the Cabinet's call for reductions in Crown outputs, specifically directing that the Ministry of Education, in consultation with Treasury and SSC, review the outputs of new education agencies such as ERO, NZQA, Special Education Services (SES), the Early Childhood Development Unit, and Quest. Carpinter also reminded the committee that the objectives of *Tomorrow's Schools* were to improve education administration by making self-governing and self-appraising schools the major focus, as well as reducing the size and powers of the centralized bureaucracy by more closely specifying its tasks. Carpinter claimed that

> During the implementation phase a number of policies were watered down, deferred or rejected. Factors encouraging this were a relitigation of policies particularly by educationalists who did not agree or did not understand the intent of the reforms, a heavy implementation programme being complete within a short timeframe, and the lack of relevant skills by principals and trustees.[27]

Turning to the conclusions of the Lough Report, Carpinter conceded that there had been a high level of resistance to the devolution (bulk funding) of teacher's salaries by both teachers and educational bureaucrats, to the extent that the working group on funding could not reach agreement. Observing that the MoE was to report to the Expenditure Committee by February 28, 1991, he cited SSC's conclusion that, "in considering all education agencies, we believe that a further downsizing of the Ministry provides the greatest potential for savings."[28] Carpinter cited a recommendation of *Tomorrow's Schools*, that SES be a freestanding, self-administering body contracted by the ministry to provide service to learning institutions. In so doing Carpinter was well aware that SES had recently advised Cabinet that making services contestable would be too difficult, but he nevertheless observed that SSC, treasury, and the minister of education, Lockwood Smith, had all argued that SES had not substantiated its own position or met the requirements of the cabinet directive. Neither had it advanced the government's policy on special education.[29] Turning to the Teacher Registration Board (TRB), the agency responsible for determining the conditions and requirements under which teachers were registered, Carpinter reiterated the government's recent decision that TRB be a small board with a majority of non-teachers to avoid professional capture. However, the problem as Carpinter saw it was that

> all the current board members and the director, with one exception, are practising or former teachers. The Board is seen as over-identifying with the professional interests of teachers and that the concerns of employers, taxpayers and potential teachers do not seem to figure as large.[30]

In early March SSC's political chief, Minister of State Services W.F. Birch, reminded the Cabinet Expenditure Control Committee that Cabinet had agreed in August 1990 that the Ministry of Education report to Cabinet Expenditure Review Committee by February 28, 1991 on how it proposed to reduce the cost of its outputs by the target date of 1992/1993. That MoE was not about to meekly capitulate to these downsizing demands, however, is reflected in Birch's unhappiness with its resulting report.

In response to the MoE report, Birch reiterated the Lough team's conclusion that, even following the implementation of the Picot-inspired reforms, a high level of resources still remained with the central administration.[31] His suspicion that the ministry was dragging the chain over implementing reduced expenditure targets was illustrated a few days later in a memo to the minister of finance, Ruth Richardson. In the memo, Birch warned that SSC was

> concerned that the Ministry is offering very limited immediate savings indeed and still has very few firm proposals for substantial downsizing. It is still talking vaguely about "flexible staffing structures" which will be achieved (by 30 June 1993) by regular internal audits, rather than indicating exactly when and how it will change the amount and content of outputs in which large numbers of staff are involved. In particular in the Operations area, as the Minister of Education has asked for. It will be most unlikely that the Minister of Education will achieve his target of a reduction of $25 million across all central Government Education agencies if the Ministry is not one of the major contributors.[32]

Birch and SSC were particularly critical of the Ministry's lack of clarity regarding the future options for both the Payroll Service and Learning Media as stand-alone commercial activities.[33] Hence, they felt, that the Ministry was "offering only vague and long-term staffing reductions and very limited concrete and immediate reductions in the costs and outputs of the Ministry." Accordingly, it should be directed to "reduce staff to 400 by 30 June 1992 or reduce the costs of outputs by 50 per cent by 31 December 1992."[34]

In March, however, MoE responded more emphatically to its critics by presenting the draft report of its review team. Chaired by the Ministry's Group Manager: Operations, Catherine Gibson, the report firmly rejected the option of selling Learning Media on the grounds that it placed the interests of users in jeopardy with regard to accountability, product quality and reliability of supply as well as being unresponsive to government priorities. Its preferred option was not privatization, but to make Learning Media a Crown Agency.[35]

The National government, however, remained steadfastly unsympathetic to these arguments. On March 12 a meeting of the Cabinet Expenditure Control Committee took place. Those present included a formidable line-up of senior National MPs including Doug Kidd, Bill Birch, Ruth Richardson, Jenny Shipley, Warren Cooper, John Luxton and Lockwood Smith, together with officials from the Prime Minister's Department, Treasury, SSC and Education. The committee began by castigating the MoE for considerably exceeding previously agreed figures for teacher's salaries and relief teachers, complaining that "over-spending of this nature severely compromises the government's fiscal management."[36] The committee agreed that Ministry should hire consultants to develop management and financial information systems leading to improved forecasting and control. The consultants were to report to the Officials Committee, which would determine the ministry's new terms of reference. SSC negotiators would attempt to ensure that the forthcoming teacher's wage round would produce net savings, meaning that zero or minimum wage adjustments were desirable, with tradeoffs in employment.[37]

A document to this effect sent to the Ministry on March 15 reiterated that the Lough committee had recommended that MoE's outputs be reduced over a two-to-three-year period from April 1990. It was pointed out that the Ministry's previous report detailing how it was planning to reduce the cost of outputs had attracted expressions of concern by Treasury, SSC, and Manatu Māori; that MoE estimates in 1990/91 were already $68 million over-budget in addition to the further expenditure increases already noted.[38] By 20 March SSC had formally rejected all three of MoE's preferred options for Learning Media, asserting that "the only logical option" was to sell the agency as a going concern.[39]

That the intensity of the struggle was taking a toll on those most involved in the difficult face-to-face negotiations that now followed is clearly revealed in the terse memos exchanged between Treasury, SSC, Social Policy, and MoE officials during March 1991. At one point David Tripp claimed to have run into a pretty heated John Gill about the revised set of ministry outputs, grumbling that the process which was to lead to a revised set of outputs was frustrated by "John's 'republican guard' in financial management accounts—the people who are used to bucket funding. There is no involvement by any corporate planning types." Clearly upset by what he regarded as MoE recalcitrance, Tripp believed that it was going to be "a long and hard road, given John's attitude."[40] For his part Gill felt that the Treasury officials were going far beyond their brief. He pointedly emphasized that "it is important that we establish the ground rules. If the end of the line is that you formally direct me to follow your instructions, as you have done with the POBOCs for *Tomorrow's Schools*, then please say so."[41] In response to this memo Treasury officials denied that Treasury was formally directing MoE to undertake a particular cut in its outputs, Greig quipping "Well, it is indeed a cheery note from our friend John."[42] Nevertheless, Gill's was a courageous stand.

Once again the irony is that, apparently unaware of the battle then raging between the various state agencies over the pace of devolution, the Lauder Report prepared by Professor Hugh Lauder of Victoria University's Education Faculty, was advancing similar conclusions to that of Treasury and SSC regarding the continuing power still accruing to the central education bureaucracy, with the culprits in this case being Treasury, SSC and the National government.[43] In this report, commissioned by the Education Sector Standing Committee of the New Zealand Council of Trade Unions, Lauder began by arguing that although the market-led tendencies in the reforms had originally been balanced by the Labour government's insistence that policies of equal opportunity should play an important role, the election of National in 1990 had seen the process of mercerization advancing rapidly.[44] Lauder alleged that the pretext of economic crisis was used to reshape education according to market principles and privatization, and to cut educational expenditure. Although the original Picot-inspired reforms were substantial enough, a second wave of reform introduced under National saw thirteen separate reviews of education. Citing the work of several contemporary educational policy researchers, Lauder concluded that "the setting up of review committees without representation from teachers, or boards of trustees or councils, gives credence to the view that an intention of the reforms was to decentralize responsibility while power remained at the centre."[45]

The view embraced by Left-liberal critics who held the Picot reforms to be a cynical camouflage for established vested interests was also being promoted by neoliberal critics outside of immediate government circles. The book, *Upgrading New Zealand's Competitive Advantage*, published in March 1991, was a near contemporary of the Lauder Report. Led by Harvard Business School Head, Professor Michael Porter, the Porter project was a neoliberal initiative that attempted to take "a deep and fundamental look at the New Zealand economy and the history, national attitudes and institutions that have shaped it."[46] The authors argued that the country's once-high living standards had faltered because industry had failed to broaden and upgrade its competitive advantage to cope with increasing international competition. The root cause lay in outdated government policies of protectionism and continuing attempts to maintain a social welfare system that prioritized wealth redistribution at the expense of innovation and skills development. Accordingly:

> The New Zealand education system has placed a higher priority on the transmission of societal values and academic training rather than on providing economically useful skills. The resulting skill base makes it difficult for us to develop higher-order competitive advantages and reinforces a focus on price competition in industry.[47]

It followed that the educational solution lay in expanding the notion of education to include economic as well as social and academic goals. These included developing appropriate standards that evaluated and motivated performance at the individual, institutional, and systemic level; teaching the skills necessary for youth to become productive members of society in an increasingly competitive world with increased focus on mathematics, technological subjects, and languages; and improving interactions with industry through curriculum development, internships, and joint research. In this way the education system as a whole and secondary schools in particular would be realigned to meet the needs of industry.[48]

Given the prevailing mood within Cabinet, it is hardly surprising that these recommendations were to have an immediate political impact. Within weeks of the Porter Project's conclusions being published, Lockwood Smith was noting the educational solutions it advanced with particular approval. In a draft thirty-seven-page document entitled "A Strategy for Investment in Education and Training," presented to the Cabinet Strategy Committee, the Minister concluded that "While investment in education and training is properly for social as well as economic reasons, the balance has swung too far in favour of non-economic objectives, and this imbalance threatens our ability to achieve both our economic and social objectives."[49] Smith went on to cite the Project's conclusion that New Zealand's education system tended to focus on social objectives rather than subjects with direct economic value such as sciences, engineering, mathematics and management. Consequently, the school curriculum at all levels displayed little regard for skill bases that would be critical to success in an increasingly technological world. Participation and performance in such areas as mathematics and science was poor, compared to other advanced nations.[50] All this could be attributed to the fact that "Our education system has a distinct academic and theoretical basis," hence

> what is wanted is a major shift in the orientation of education towards the flexibility and adaptability of educational institutions. Restructuring is already in progress and more is to come. The necessary changes are substantial and, as we have seen already with bulk funding in schools, will be strongly resisted by at least some of those with a vested interest in the status quo. Restructuring must not involve a rejection of social and cultural values and their place in the curriculum: these must be retained. What is required is a significant reorientation of emphasis to give at least equal weight to New Zealand's economic requirements.[51]

Smith's views confirm the suspicions of both Left-liberals and neoliberals that, although the Picot-inspired reforms did not directly address curriculum or assessment issues at any level of compulsory schooling, they nevertheless provided a political context for more frequent and direct government intervention in the name of "the national interest." Indeed, a key outcome of the administrative changes initiated by Picot Report and *Tomorrows' Schools* was a simplified structure of administration that enabled the forging of much tighter links between policymakers and schools. Over a century before, during the debate over the introduction of New Zealand's first national system of education in 1877, some MPs had warned prophetically that the introduction of a two-tiered system would ultimately mean that there was no cushion between individual schools and the priorities set by governments. Outlining the achievements of educational reforms since 1989, a 1993 MoE document merely confirmed the truth of this prediction, asserting that the reforms had brought about "a clear policy framework for developing and implementing the national curriculum and local school programmes." According to the Ministry, the reforms had also encouraged more rapid development of curriculum statements; brought additional funding for curriculum implementation; enhanced provision for Māori in through the development of curriculum statements and supporting documents in the Māori language; increased local community participation, delivered more flexibility and autonomy for schools to determine their own curriculum needs, produced an increasing number of curriculum development programs for principals and teachers; and stimulated the production of learning resources to support the curriculum.

Both neoliberal and Left-liberal goals were thus to be represented. Indeed, curriculum and assessment initiatives, post-1989 display a decidedly mixed ideological parentage. The release of the *New Zealand Curriculum Framework* in 1993, for instance, reflected the contemporary neoliberal ideals of Treasury, SSC, and the Porter project in its assertion that

> New Zealand today faces many significant challenges. If we wish to progress as a nation, and to enjoy a healthy prosperity in today's and tomorrow's

competitive world economy, our education system must adapt to meet these challenges. We need a leaning environment which enables all out children to attain high standards and develop appropriate personal qualities. As we move towards the twenty-first century, with all the rapid technological change which is taking place, we need a work-force which is increasingly highly skilled and adaptable, and which has an international and multicultural perspective.[52]

These goals accorded well with the government's educational blueprint, *Policy '93*. In this document, Smith made it clear that education was "the bridge that links the government's economic and social goals, and that it has the potential to sustain and accelerate economic growth."[53] As Lee and Hill have emphasized, however, core curricula can only shift from rhetoric to reality when they gain firm support from students, teachers, parents, and employers.[54] The shift toward emphasizing economic goals exemplified in the draft curriculum framework document, for instance, encountered considerable resistance from teachers and many parents, indicating that such support was by no means assured. Submissions critical of the draft complained of an undue emphasis on competition and market pressures, a somewhat narrow view of skills development, a perceived shift toward greater conformity, a tokenistic approach to Māori values, a failure to explicitly address equity issues, and a neglect of learners with special needs.[55]

While the pendulum had swung strongly toward economic imperatives in the curriculum, however, it had by no means swung as decisively as some critics assumed. The final *NZCF* document also encapsulated the historical assaults on the traditional secondary school curriculum trends over the previous forty years as epitomized in the conclusions of the numerous curriculum documents since the Thomas Report. The NZCF acknowledged the influence of the comprehensive reviews of curriculum and assessment:

> The reviews sought a more equitable curriculum, particularly for those who were found to be disadvantaged by the existing system, such as girls. Māori students, Pacific Islands students and students with different abilities and disabilities. The reviews acknowledged the significance of the Treaty of Waitangi and its implications for New Zealand society, according particular values to teo reo and nga Tikanga Māori, They recommended an increased emphasis on culture and heritage, to reflect a growing awareness of New Zealand society and its multicultural composition. Outlining changes in New Zealand's economy and society, the NZCF pointed to demographic changes, the emergence of cultural and gender issues, the increase in women's participation in the labour market.[56]

Subsequent documents such as the 1994 draft social studies curriculum statement were to awkwardly balance bicultural, social meliorist and social constructivist aims with the National government's emphasis on competition and skills-development.[57]

New assessment procedures displayed a similarly mixed ideological heritage. Clearly evident was a swing toward a greater recognition of national economic priorities in a rapidly changing global economy, a continuing trend that traced its origins in back to the crises of the late 1970s. Evident too was the generic influence of repeated critiques of secondary school methods of assessment since Phoebe Meikle's time. In November 1992, NZQA released National Qualifications Framework. *Learning to Learn*, an introductory booklet to the new Framework. This drew heavily on a diverse body of criticism that had characterized so much of the long-standing debate over postwar secondary education, culminating in the many submissions to Marshall's Committee of Inquiry into Curriculum, Assessment, and Qualifications in Forms 1–7. Under the heading, "Why the present system needs changing," it was argued that technological advances and changing economic conditions meant that the modern worker would have to change direction many times in his/her working life.

To some extent this was simply a contemporary restatement of the whole drive for more relevance in the secondary school curriculum that had found early expression in PPTA booklets such as *Curriculum in Change* some twenty years before. Postwar criticism of secondary school curriculum rigidities were clearly evident in *Curriculum in Change*'s view that it was a tragedy that so many New Zealanders had given up on qualifications, "discouraged by a system with too many barriers—a system which has effectively determined winners and losers".[58] However, the booklet's claim that successful industries and enterprises needed multi-skilled, adaptable workers was not too dissimilar to the much later Porter project's educational ideal of an education system responsive primarily to the needs of the economy.

At the end of National's second term, with a general election looming, the government released a White Paper evocatively entitled, *The National Qualifications Framework of the Future*. In his foreword to this document the minister of tertiary education, Max Bradford, both summed up National's hopes for a seamless education system, while epitomizing once again its mixed ideological parentage. Bradford argued that the new broader qualifications framework that had been introduced by the government was "essential for New Zealand's bright future." A key challenge facing the education system was to ensure that the qualifications were up to date and relevant. For this reason, recent reforms to qualifications policy, including the establishment of NZQA, had been designed to make qualifications more transparent and explicit about what people had learned. The reforms were also designed to involve industry much more closely in the development of qualifications. There was, moreover, a need to establish a broader range of subjects and pathways desired by the considerably more diverse student body that was the direct result of sharply increased participation in senior secondary education. Finally, there was the challenge of new subjects such as computer science, which could not be easily structured into the existing system of assessment.[59]

As we have seen, these ideals had much in common with those that had been advanced both the department of education and by PPTA for many years. National's vision of education, however, was to be summed up in *Education for the 21st Century*, a draft of which was released in July 1993 followed by the definitive document in 1994. In his foreword, Smith stated that it was intended to furnish a strategic plan that reflected his confidence that "a consensus has been achieved on what we need for out education system as we approach the twenty- first century."[60]

Curriculum and assessment were not the only areas in which National sought to significantly alter both the aims and content of secondary education. Throughout 1991 pressure had grown for a further strengthening of the devolved system through the encouragement of total bulk funding rather than the two-component bulk grant outlined in *Tomorrow's Schools*. In early 1991, an unpublished SSC paper recorded that the government had two objectives that were not necessarily compatible with one another: to implement bulk funding successfully at minimum cost, and to make savings within Vote Education. SSC conceded that the government also had to consider current resistance to bulk funding that, in its view, had been exacerbated by union campaigning. Noting that SSC had previously supported compulsion on the grounds of least risk, the paper noted that voluntary "opting-in" aimed to prove bulk funding could work and might break down resistance within schools but added the caution that it might then be difficult to make it compulsory as a next step.[61]

Birch and SSC had been particularly concerned by the results of a survey by NZCER researcher, Cathy Wylie, suggesting that a high proportion of BoT members felt overworked and overwhelmed, though she had observed that the problems seemed less evident in secondary schools which had traditionally operated with boards of governors. Moreover in late 1990 the New Zealand School Trustees Association (NZSTA) had instigated a survey of rural schools that was carried out by a special review team set up within the Ministry of Education in early 1991. Described as "the most comprehensive recent commentary on rural education in New Zealand" the resulting report edited by Colleen Pilgrim, the Director of the School Trustee's Association (STD) claimed to present "a uniquely lay and parent perspective."[62] Although broadly supportive of the goals enunciated by the Picot Report and *Tomorrows' Schools*, it also concurred with Wylie's conclusions about the specific problems faced by rural schools.

In response to these findings, SSC's three regional assistant commissioners in Wellington, Auckland and Christchurch carried out a rapid survey of a cross-section of schools in their areas, resolving to go "directly to trustees, rather than relying on hearsay or filtered reports for their information."[63] Although they noted particular problems in rural schools, a need for better BoT management skills, and the frustrations schools experienced in the continual paper war with often conflicting authorities, they nevertheless concluded that schools "were pretty flush with money," that "everyone is positive on *Tomorrow's Schools* and while there are frustrations, none would go back to the previous system."[64]

Lockwood Smith, however, appears to have become firmly convinced that the emerging bulk funding controversy was not only central to the success of his own tenure as Minister, but would also be "make-or- break," for the future of the whole *Tomorrows' Schools* initiative. To STA president, Graye Shattky, Smith expressed his concern that if the process of self managing schools now came to a halt now, "we will be left in mid-stream."[65] His strategy, therefore, was one of better resource management of resources, coupled with more power for both parents and BoTs; aims with which Shattky was in accord. Shattky felt, however, that there was no use trying to debunk the statistics about support for bulk funding. Rather, there was a need to look at strategies designed to educate BoTs in their task. Like Smith, Shattky was convinced that "if bulk funding fails, the whole system fails."[66] Smith was feeling the political pressure. Given Shattky's plea to him for support ("we need to see you arming us"), the Minister hoped to get an Education Reform Bill in front of a Select Committee over the next two weeks, with everything to be completed by June 30. To Shattky, he intimated that he was developing national education guidelines because he wanted charters to be ore focused on standards—hence he intended changing the law to require ERO to review against national guidelines. The informal meeting concluded with a resolution to work together.[67]

Implementing these goals, however, was to prove increasingly difficult. In opposition, Labour had been joined by the breakaway party, New Labour. Both inside and outside House New Labour MPs, led by the member for Sydenham, Jim Anderton, were increasingly vocal in their denunciation of what the party regarded as blatant Treasury interference in education policy. They had plenty of ammunition. In April, Smith released the draft of an Education Amendment Bill permitting secondary schools to abolish existing zoning restrictions. Where overcrowding occurred, secondary schools would be able to devise their own enrolment schemes. The bill also removed the requirement for Community Education Forums that *Tomorrows' Schools* had envisaged as "the official voice of the community on educational matters."[68] The removal of zoning restrictions in particular, proved controversial. Just two years later the 1993 ECD report, *School: A Matter of Choice*, recommended interventions to create choices for groups who were currently educationally underserved, a further diversification in the range of educational suppliers, and increased choice options. Responding to these recommendations in September 1994, Labour's Margaret Austin pressed the Minister to "concede that removal of the right of access to the neighborhood school in 1991 has resulted in blatant selection of some students, and not choice for the parents, to the extent that some schools are now behaving as Government-funded private schools."[69]

Most controversial of all, was an exclusive front-page article in the major Wellington daily, the *Evening Post*, which claimed sensationally that Treasury had submitted a confidential paper to the government advocating that schools be sold, lock, stock and barrel, to BoTs.[70] Although Treasury was quick to publicly deny the claim,[71] the damage had been done. Anderton was particularly concerned at what he regarded as "the harebrained policy advice provided to government by Treasury zealots." In a scathing press release he hinted at future plans to privatize schools and transfer ownership to BoTs which would give trustees "the power to sell the property and go for a long holiday in the South of France, to turn the school buildings into a casino and train the pupils to spin the roulette wheel as part of their maths training, or to go bankrupt."[72] Although this release does not appear to have appeared in its entirety in the national press, a subsequent *Evening Post* editorial condemned the "harebrained idea" that BoTs should have ownership of schools transferred to them as "a signal" of current thinking that the state only had a minimal involvement in societal responsibilities.[73]

The direct impact on the public of the war of words being conducted through the media is uncertain. SSC claimed to be "pleasantly surprised at the level of support for bulk funding and the self managing school among the secondary principals we have spoken to recently." However, SSC conceded that the opt-in rate for bulk funding could be quite modest, with few, if any, primary schools likely to opt in, even though although a "reasonable number" of secondary schools might.[74] Moreover, it was concerned that the government was fiscally vulnerable by currently giving BoTs "employer-power" without being responsible for the decisions they might make. SSC was also increasingly doubtful about the viability of voluntary bulk funding, which would not in its view resolve the wider question of controlling expenditure on teacher's salaries.[75]

The Education Reform Bill was not to have its second and crucial reading until December. It stopped short of making bulk funding compulsory. Smith was later to claim that, "the teachers' unions had found it unacceptable and we turned that legislation round and made it that it was not enforced nationwide."[76] Despite the volte-face, Smith stated in moving for the Bill's acceptance that it gave effect to key government education policies outlined in the government's election manifesto. These included greater freedom from the compulsory charter framework, and the relaxations of current restrictions on the size and composition of BoTs to allow for the nomination of individuals who were not parents of students attending the school:

> I guess one could say that it restores to education the vision of the Picot task force, because many people felt that there was much good sense in what the Picot task force proposed for the New Zealand school system. Many people feel that when the previous Government implemented its *Tomorrows' Schools'* reforms that grew out of the Picot task force proposals much of the vision, the flexibility and the freedom proposed by the Picot task force was lost. It was trampled under the previous Government's social agenda.
>
> The Bill restores to our school system that freedom of vision; that vision to enable our schools to be responsive to their communities, to be flexible, and to respond to a world that is ever-changing, increasingly rapidly—it is nothing imposed on them by a Government, but the reality of the modern world is that it is rapidly changing. The rate of that change is increasing and schools must have the flexibility not only to respond to that change but to use change to benefit their students.[77]

Accordingly, the new legislation proposed substantial changes to existing education guidelines as laid down in compulsory school charters. Given that schools were "mostly about learning," it was of particular concern to National that the equity principles centering on equal educational opportunity and the Treaty of Waitangi accounted for three out of four of the compulsory charter framework's guiding principles, leaving only one guiding principle headed "curriculum." Hence, new national education guidelines were to focus on achievement goals—skills, understanding and knowledge relevant for "the modern world." They would include national curriculum statements with clear essential learning areas, skills and objectives, together with new requirements for schools to report on student progress.[78] In addition, the school leaving age was to be raised to 16, given "the 'high tech' and competitive world in which people's skills and education (were) critical to their ability to compete." Smith sought to reassure potential critics, however, that the government was not imposing provisions on schools as "schools will not have to change one word in their charters if they do not want to do so."[79]

During the Education Reform Bill's second reading, Smith also revealed that the government was developing a new training opportunities program together with a National Certificate qualification based on units of learning which provided opportunities for students "disillusioned with the traditional school fare." The new qualification would span both the senior secondary school years and the polytechnics, permitting students to take units of learning that were of interest to them.[80]

Not surprisingly, despite Smith's reassurances, Labour Opposition members saw the bill as little more than a blatant attempt to over-ride the intent of *Tomorrow's Schools'* democratic provisions and equity safeguards. Margaret Austin, Labour's educational spokesperson pointed to the final removal of requirements for the minimum staffing of schools—"part of the combined State Services Commission and Treasury agenda." The Minister, she claimed, was desperate to introduce bulk funding by offering inducements to schools to join the scheme. Hence, "the bill was ideologically driven."

> The changes made to the constitution of boards of trustees...mark a real shift in the balance away from parents having the prime responsibility for decision-making. The weight of the submissions presented to the select committee stressed over and over again the importance of parent representation on boards. Ministry officials warned against providing organisations independent of parents with placers on boards, stating that it was in conflict with *Tomorrow's Schools* and, indeed, the Government's own policies.[81] Moreover, the Minister had failed to acknowledge that when that provision is combined with allowing anyone to be nominated for a position on the board—even of he or she is elected only by parents—it is clear that the whole balance of power is shifted away from parents, and *Tomorrows' Schools* is eroded significantly. All of those non-parents will have the same status as parents. They will not only make decisions but will also have the ability to co-opt other people to the board.[82]

The Reforms Cemented

By this point it was evident that any government-imposed changes to existing educational structures ran the risk of Opposition parties invoking the spirit of Picot. The general election in November 1999 saw Labour sweep back into power. It was noteworthy that, like National before them, they did not seek to overturn the educational reforms. Indeed, they continued and even accelerated the extent of direct government involvement in curriculum, assessment and the maintenance of academic standards. Once again this was justified in terms of the national interest within the context of a global economy, but now with a greater emphasis on equity issues, including Treaty of Waitangi obligations than had been evident under National. While in Opposition, Labour had also criticized the government's failure to raise academic standards in key subjects. In May 1999 for instance, Labour's spokesperson for Education, Trevor Mallard, highlighted the lack of progress in the performance of year 5 and 9 pupils evidenced in the preliminary trend results of the repeat third international mathematics and science study carried out in 1998.[83] Hence it was not surprising that as Labour's first Minister of Education, Mallard firmly rejected what he called "bulk funding and 'cherry picking'—the failed policies of the 1990s," in favor of a comprehensive strategy that included literacy leadership for principals, whole school professional development, new assessment tools for teachers, and the production of enhanced learning materials.[84] In 2000, Labour introduced a large scale numeracy professional development project which by 2006 was to involve more than 1600 schools. Teachers were to be given new diagnostic assessment tools such as asTTle (Assessment Tools for Teaching and Learning), claimed by Mallard's successor to the education portfolio, Steve Maharey, to be an approach that ensured that "teachers, parents, and students understand where students went wrong and how to improve their performance." Thus the focus was to be "on achieving the full potential of a student rather than on centrally imposed minimal standards."[85] Additional measures designed to lift the performance of Māori students such as Te Kotahitanga were introduced from 2003.

In its first term of office Labour also introduced a National Certificate of Educational Achievement (NCEA). This built on the assessment reforms mooted under National in addition, it may be recalled, to reflecting the tenor of many submissions to the 1984 Committee of Inquiry into Curriculum, Assessment, and Qualifications in Forms 5–7. Accordingly in 2002 NCEA level 1 replaced the existing School Certificate level, with level 2 and 3 following in 2003 and 2004 respectively. NCEA also reflected the previous decade of curriculum reform in that it aimed to provide a bridge between school, workplace, and so-called lifelong learning, for long a common goal of teacher, employer and business organizations seeking reforms to the existing secondary school assessment regime. Credits gained from NCEA and its battery of assessments at successive levels across every subject during any given year of a student's secondary education program could be transferred through to many of the other 800 or so qualifications within the National Qualifications Framework.[86]

An information sheet aimed at secondary school students and parents embarking on NZCEA for the first time observed that the "Record of Learning" supplied to each student would show students, parents, teachers and employers how well individuals performed in each of the separate skills and knowledge in every course undertaken. Hence, the document claimed that "when your NCEA results arrive, you'll see a full picture of your achievement."

> Ever wondered why your teacher gives you 8 out of 10 for a piece of writing? Or what 74% meant in School Certificate? Often those marks are about rankings—how your work compared with others. And a mark of 74% didn't show which bits of a subject you were good at. Even if you got 36% for School Certificate you were probably good at something but you got no credit for it. No one knew what your strengths were.[87]

Once again, behind this informational document lay the historical assessment concerns of both educators and employers over the previous fifty years. Likewise, the introduction of NCEA was to stir a public debate all too familiar to historians of New Zealand secondary education. In February 2005 it was revealed that the 2004 NCEA scholarship level examination results showed huge discrepancies between subjects. To give one example, just nine out of 1000 biology scholarship candidates passed, compared with 300 out of 900 English examination students. As had so often happened in the past, this revelation prompted a barrage of criticism from press and public. A typical reaction came from the *Dominion Post*, which argued that some might well see accountability for the debacle ending jointly with both Mallard and NZQA's then-chief executive, Karen Van Rooyen.[88] Despite considerable political pressure, however, the government, although welcoming a proposed SSC review into the situation, was quick to scotch any suggestion that NCEA would be abolished as a result. The *Dominion Post* cited Associate Minister of Education, David Benson-Pope's rejoinder that, "there ain't any alternative and we're not looking at any alternative...end of conversation."[89] A feature article in the same issue of the newspaper observed that far from wanting to scrap NCEA, PPTA wanted more funding and resources to implement the system properly, while mentioning the fears of some employers that NCEA might "dumb down" the secondary school system.[90]

Ironically, although some commentators continue to critique NCEA as the product of the same neoliberal ideology that is claimed to have originally driven the Picot Report and *Tomorrows' Schools* and others denounce the qualification as encouraging mediocrity, NCEA seems to still enjoy continuing support from many secondary school teachers. A commentator in the *Otago Daily Times*, for instance, observed that it was noticeable that the great majority of secondary school teachers were not "... rising up against NCEA."

> Why not? Because most of us can see that, while not perfect, it is a big step forward from the muddle of school certificate, sixth form certificate, bursary and scholarship awards it replaces. Pupils today find it hard to believe that not so long ago, those who sat their end-of-year examinations had their actual marks scaled to ensure half of them failed. They scoff at the idea that we used to assess practical skills, such as public speaking, not by grading an actual speech but by sending pupils into an examination to write about a speech they may or may not have actually given. It was also ludicrous that, in some subject, a few schools, which had been part of an internal assessment trial for school certificate in the past, were allowed to carry on internally assessing their pupil's work, but all other schools had to use the examination. Many pupils were unable to motivate and organise themselves to work towards a goal which they saw as distant, and tried to fit a year's work into the final few weeks of cramming. Others got to the end of the year and realised, too late, so didn't bother trying.[91]

This verdict aptly sums up many of the concerns about the former School Certificate-University Entrance-Bursary system that had been articulated by many educators throughout the postwar era, including those of Meikle, the PPTA, and numerous academic critics. Thus, in many ways the whole NCEA debate continues to illustrate both long-standing and ongoing issues over secondary school level assessment, while encapsulating the essential ambiguity of the nation's reaction to the legacy of the wider educational reforms initiated by the Picot Report.

Conclusion

A Real Say or a National Morality Play? The Road to Radical Reform in Retrospect

In February 1990, some nine months prior to a general election that saw National sweep into power, Elizabeth Tennet, the Labour member for Island Bay, sought to justify the wave of radical educational reforms ushered in by the Picot Report and *Tomorrow's Schools*. The biggest problem, Tennet maintained, had been the economy, exacerbated by many failed attempts to insulate New Zealand from events elsewhere in the world. While other countries had reskilled, emphasizing education and skills training, New Zealand had done nothing. The result was that during the previous thirty years the nation's standard of living had dropped from third to seventeenth in the OECD.[1] Before 1984, she claimed, New Zealand had falsely prided itself on having a well-educated work force, even though participation in skills retraining by young people over fifteen had remained one of the lowest in the Western World. Fortunately, however

> the *Tomorrow's Schools* reforms have now given parents a real say in their children's education. The administrative problems and fears about funding the implementation have unfortunately clouded the reforms. However, the importance of decentralisation should not be downplayed. Parents will now demand better education from schools, and they will demand education for their children that promotes job prospects, and a balanced social individual.[2]

Some nine years later however, the Massey University College of Education Policy Response Group's unofficial briefing paper for the APEC Summit concluded that the purpose of the *Tomorrow's Schools* reforms had proved "frustratingly difficult to identify." Worse, there was

> scant evidence of any tangible improvements to teaching and learning as a result of site-based management. Despite a decade of market liberal reforms, continual classroom change and constantly shifting central government priorities there is little evidence of success...a feature of the reforms has been a demand for clearly specified objectives, objective measurements of results and strict accountability. In these terms the zealous reformers of the 1990s have failed to measure up to their own professed standards.[3]

The Evidence for Success

These two widely differing assessments of the success of the reforms and their underlying motives once again highlight the persistence of polarized views on their impact. Even today there is little agreement over exactly what was achieved, whether the overall impact has been beneficial or not, or even what might count as success. Reform supporters, such as Tenet, have argued that increased parental involvement in schools in itself constitutes evidence that the reforms have not only been successful in their aim, but have also struck a responsive chord in the wider New Zealand society. Although there has been no corresponding data for secondary schools, Cathy Wylie's annual surveys of the impact of *Tomorrows' Schools* on primary and intermediate schools provide some indicative information on BoTs. Some four years after implementation, Wylie concluded that while the workload of principals, teachers, and BoTs had soared, "both volunteers and professionals have (largely) worked together in harmony rather than conflict."[4] Moreover, the oft-expressed fear that trustees would "patrol classrooms and curriculum with rigid and outdated beliefs of what should be taught, and how" had not been borne out.[5] Rather, schools seemed generally to be more aware of their relationship with parents. She observed, however, that some parents were becoming more selective with regard to school choice, while some schools were more selective in taking students. The surveys Wylie conducted also revealed that existing resource gaps between schools appeared to be widening rather than closing.[6] The best resourced schools tended to be in middle-class areas with a high proportion of European students while the least resourced schools tended to be in working class areas with significant Māori rolls. This, Wylie concluded, was a serious issue given that one of the aims of the reforms had been to overcome existing disparities.

Further evidence that educational disparities have remained a problem despite the reforms appeared to be confirmed by a series of brief reports in the *New Zealand Herald*. Largely on the basis of interviews with a number of academics, principals, and teachers, the reports critically examined how well the system had fared after ten years of *Tomorrow's Schools*.[7] The findings were presented as a report card under eight headings: zoning, marketing, Māori, principals, trustees, ERO, Funding, and standards. Pointing out that parental choice had not happened for all, the report observed that the best schools selected students they wished to admit. The most successful schools tended to market heavily, but advertising was seen to cut into limited funds, with many teachers resenting both the time and cost involved in marketing. Māori were still disadvantaged, with some 40 percent of Māori leaving school with no formal qualifications in addition to featuring disproportionately in suspensions/expulsions. School principals were now working longer hours and reported spending more time on administration than on leadership. Disadvantaged schools tended to have disadvantaged BoTs. The reforms had widened the gap between rich and poor schools, with many local communities digging deeper into their own resources to provide support for schools.[8] The report was not all bad news, however. It was concluded that most parents seemed happy with the new system. Citing a recent Herald-Digipoll survey of 400 parents, the report claimed that 85 percent of those with children currently at school were satisfied with their schooling, while those whose children had recently left school were even more satisfied (88.5%).[9]

Overall, the report revealed the existence of a sharp gap between parents and educational professionals regarding the efficacy of the reforms. Returning to the original question of whether the reforms had delivered on what they promised—particularly as regards choice, power and equity, the *Herald* concluded that "in the first two areas particularly, there have been improvements but there are also failures which signpost the challenges ahead." Asked to indicate what these challenges might be, Rangitoto College principal Allan Peachey stated that he would like to see successful schools given greater autonomy, and for the state to provide different models of management and leadership "for those communities that need intensive work to improve their schools." Asked the same question, Brian Picot conceded that were he to have his time over again, "he would give greater emphasis to the need for administrative assistance for schools and the advantages of small schools clustering together."[10]

The *Herald's verdict* on the reforms after a decade of operation then, seemed to be both confirming some of the best hopes of supporters and some of the worst fears of critics. In an article for the *Dominion Post* marking the twentieth anniversary of the reforms, Wylie again turned her attention to the question of how well parents were managing their responsibilities as BoT members. Conceding that there had been little research into exactly how BoTs contributed to the performance of their schools, she suggested that at least, "sitting round a board table gives school professionals and members of their community the chance to learn from each other, and to forge links and shared values that might be harder to build otherwise."[11] Citing NZCER surveys of primary schools (since 1989), and of secondary schools (since 2003), Wylie observed that about 15 percent of secondary BoTs at any one time were experiencing difficulties, with some 70 percent of boards reporting gaps in expertise in areas such as legal skills and strategic planning. The situation appeared most acute for schools in lower socioeconomic areas. She also pointed out that there were tensions in the principal-BoT relationship in 10–15 percent of schools at any one time, and that up to a quarter of boards felt they had insufficient information upon which to base decisions. Wylie conceded, however that there was "no sign of a groundswell of parental dissatisfaction with boards, or of serious scrapping among trustees or worth their schools," hence "the system (was not) about to fall over."[12]

In fact, the 2006 survey and earlier ones appeared to provide a more than satisfactory report card concerning working relationships at the local level. Working relations between secondary school trustees were rated "good" or "very good" by 92 percent of respondents. Three-quarters reported a very good relationship with their principal, a further 18 percent rated it good. Principals on the whole concurred. Almost all boards—91 percent—had consulted their community in the past 12 months.[13]

Given these results, Wylie did not see an obvious alternative to the BoT system, although she did believe that it was time to address the shortfalls she had identified. Her suggested remedies included MoE-funded development and dissemination of material on strategic planning that could be used by BoTs, a requirement to include in the principal selection process a representative from a local team of accredited education professionals contracted by the MoE, and the setting up of a formal disputes resolution process for parents and students.[14]

All this seemed to suggest that the new structures, after an initially restless period, had finally bedded down—at least as far as working relationships at school level were concerned. A recent paper presented to the 2008 PPTA Annual Conference by its Executive, however, was emphatic in labeling the reforms ushered in by the Picot Taskforce and *Tomorrow's Schools* as "a mistake," suggesting that the teacher's unions were one influential grouping that remained largely unreconciled to the reforms.[15] Citing a 2007 paper by Wiley that compared the New Zealand reforms with those of Edmonton, Alberta, the Executive observed that many Canadians

Wiley had spoken with had been against every school having their own boards on the grounds that such a system was inefficient, and risked framing schools in the parental interest rather than that of the wider community.[16] Claiming that the title of the Picot Report, *Administering for Excellence*, betrayed its origins in the public sector management theories popular at the time, the paper concluded that a major outcome "had been a systematic polarisation of schools along ethnic and socioeconomic lines, a result not so much of white flight as of middle-class flight."[17] Moreover, given the grim prospect of declining secondary school rolls, the scene was set for more competition, more failing schools, and a fall in the quality of educational provision.[18]

Although equity issues and the continuing alienation of the teacher unions continue to be strongly evident, other critics point to the continuing concentration of resources within the centralized state bureaucracy. In November 2007, a *Dominion Post* article claimed that the MoE now employed no fewer than 2, 554 staff.[19] By March 2008 the size of the state bureaucracy overall had become a major election issue. Questioning the dramatic growth in the numbers employed in the state sector, one newspaper editorial argued that it was "hard to see the justification for the 40 percent growth in the number of people employed by the various education bureaucracies over the past eight years."[20] At the time of writing in late November, a newly elected National Government under Prime Minister John Key was promising savings and cost-efficiencies, although to date there has been no hint of any proposal to change existing structures.

Some critics also allege that, far from extending public participation in education, the Picot Report and *Tomorrow's Schools* both legitimated and formalized an evolving model of educational decision-making based largely around sectional interests within the new education bureaucracy. For example in early 1992, an internal MoE document identifying key issues within the National Curriculum Draft for development, listed a number of areas for further consideration including Te Reo Māori and Tikanga Māori; gender/girls and women; students with specific learning needs; Pacific Islanders and multicultural; the place of attitudes and values; the spiritual dimension; and early childhood education—as well as all the new so-called learning areas. Ministry officials and "interested individuals" were invited to put their name beside those teams they wished to join.[21] A confidential internally circulated critique of ministry policy contended that such approaches reflected a highly problematical team-based matrix approach to decision making that was very much part of the ministry's then operating style. Within the Policy Division, for example, ad hoc teams of analysts with an appointed convener typically undertook development work. Those selected for team tasks usually had similar educational backgrounds, having been chosen on the basis of both their expressed interest in a particular project, and a judicious balancing of internal interest groups within the division. The result was that some issues, such as the Treaty of Waitangi and gender issues, were protected from rigorous debate by an all-embracing political correctness. Quality analysis was poor and there was little direct accountability.[22]

This situation appears to confirm the suspicions of the Lough Report writers, Treasury and SSC that the bureaucracy had indeed effectively blunted the intentions of the Picot Taskforce. It could be argued, however, that broadly similar approaches to decision making had also characterized the former Department of Education, especially toward the end of its existence. Indeed, the apparent continuities of style and philosophy within state bureaucracies often make it difficult to accurately assess the impact of the 1980s reforms across the education sector as a whole. Neither the Picot Report nor *Tomorrow's Schools*, for instance, dealt expressly with curriculum and assessment issues. Moreover, as we have previously noted, much of the ideology and assumptions that now underpin the secondary school curriculum and assessment procedures were common to various reports, inquiries and critiques of secondary education throughout the post–World War Two era. Thus, changes aimed at putting secondary schools more in touch with the needs of their local communities and the requirements of modern society, including the development of a more competitive economy, together with attempts to meet the needs of a diverse student body, had already been partially implemented well prior to the Picot Report. These changes would doubtless have continued even if the taskforce had never met.

All this highlights the difficulties involved in assessing the relative impact of particular ideologies across the broader spectrum of secondary education. In her recent timely analysis of the influence of PPTA on curriculum and assessment, Judie Alison has dismissed the claim, widely subscribed to in educational circles, that the reforms were solely neoliberal in nature. Rather, she sees the struggle over school qualifications as a struggle between conflicting discourses. In Alison's view, it is unwise to categorically reject recent qualifications changes as amounting to a neoliberal takeover of schooling in New Zealand.[23]

Critics of the reforms are arguably on safer ground in claiming that the reforms considerably accelerated the pace and direction of secondary education curriculum change. From 1990, a newly elected National government initiated a package of curriculum changes in response to the eight-level *New Zealand Curriculum Framework* (NZCF), that provided a structural template of seven "essential learning areas" (Language and Languages, Mathematics, Science, Technology, Social Studies, The Arts, and Health and Physical Well-being) and eight "essential skills" (Communication Skills, Numeracy Skills, Information Skills, Problem-solving Skills, Self-management and Competitive Skills, Social and Cooperative Skills, Physical Skills, and Work and Study Skills). This Framework applied to all students in primary (Years 1–8) and post-primary (Years 9–13) schools.

One outcome of this was a movement away from the Thomas Report's emphasis on a "generous and well-balanced education," emphasizing personal and social as well as economic goals, in favor of a new focus on a "training culture" that favored economic and competitive imperatives and outputs.[24] The pendulum did not swing completely, however, with the result that these two arguably incompatible objectives remain oddly juxtaposed in many current curriculum statements. Furthermore, in the first years of the new millennium the Labour government under Prime Minister Helen Clark continued to endorse the contradictory aims of the NZCF, with little acknowledgement that goals such as competition and cooperation, knowledge and skills, are in any way antithetical.[25] Official documents and pronouncements thus actively promote aspects of a utilitarian educational philosophy while asserting the intrinsic benefits of education in the Beeby tradition.[26]

One way in which the NZCF sharply differs from England's National Curriculum is in its embracement of an essentially constructivist approach to learning. In some ways it might even be suggested that NZCF represents a belated response to developments in the immediate pre-Picot decade quite as much as it reflects the demands of post-Picot globalization. As we have seen, the 1987 *Curriculum Review* was but the last in a long list of calls to modernize the secondary school curriculum in order to better to suit the changing economic and social climate. Both Left-liberal and Right-of-Center critics have observed that this was to be achieved through a paradoxical process by which governments assumed a more direct role in monitoring academic achievement and national priorities (however these were defined), while promising schools and their communities greater participation to ensure that both curricula and pedagogies reflected local interests.

This latter paradox was arguably intensified following the release of *Learning for Life 11*. Introduced in August 1989, this document included a clause empowering the Minister of Education to intervene in curriculum design on the grounds of safeguarding the national interests. However, the subsequent replacement of the former CDU by a new Curriculum Contracts Division signaled that curriculum design was henceforth to be a contestable process, with development contracts to be let to successful external bidders. The contractor selected was to provide for community consultation and meet a series of mutually acceptable milestones leading to the delivery of a draft statement. Once the ministry received the statement, the document was then to be released for a period in which public submissions were permitted. Depending on the submissions, the document might either be approved by the Minister, be subject to modification or, in some cases, redrafted by a third party selected by MoE.[27] In the case of the social studies curriculum statement, this process resulted in no fewer than three drafts, with the last and considerably abbreviated version presumably intended to avoid further controversy by resorting to relatively general prescriptions leaving schools (presumably in consultation with their communities), to fill in the details, especially where stated national goals might be seen to conflict with community attitudes.[28]

Much the same mixed conclusions that can be drawn about the secondary curriculum could be applied equally to assessment. As we have seen, the number of students who leave secondary school with few or no qualifications has been a more or less continual focus for critics of secondary schools of various ideological persuasions throughout the entire post–World War Two era. Predictably, much of the sentiment behind the Picot reforms centered on the high failure rate, particularly amongst the less academic, Māori and other groups of students who, ironically enough, were perceived to be most disadvantaged under the old system. But in late March 2008, twenty years *after* the release of *Tomorrow's Schools*, figures were released that showed that 38 percent of year eleven students failed to pass NCEA level one in 2007.[29] Inequities in outcome for Māori students persisted, but the former concerns over girl's achievement were now replaced with worries about the achievement of boys, given that in the early twentieth century girls continued to outperform boys at every level of the senior secondary school. There seems no reliable way of knowing what influence, if any, the reforms may have had on the current gender gap.

Suffice it to say that despite a major overhaul of the secondary assessment system, public dissatisfaction with both its procedures and outcomes appears to remain as strong today as it was prior to the reforms. At the time of writing, however, there continues to be an expansion of internal assessment at the expense of external examinations, despite the recent furor over NCEA that resulted in some secondary schools ditching the government's flagship secondary school qualification due to parental dissatisfaction. Some critics, it should be noted, have also claimed that schools had deliberately attempted to boost their pass rates by loading their programs with unit standards that were less rigorously assessed.[30]

Given all the contradictions noted above, it is perhaps not surprising that there has been a renewed assault on the comprehensive ideal in the national press. In 2005 an editorial in the *New Zealand Herald* alleged that "the state system is designed to produce a broadly similar range of ability in all schools." The editorial went on to recommend that specialized schools be created deliberately to provide for different types of students.[31] In 2007 a *New Zealand Herald* article claimed that "as many as one in five pupils in the system is failing," indicating that there was "a large group at the bottom who are not succeeding."[32] A former president of the Mangere (Primary) Principals' Association, Keith Gayford, claimed that the outdated secondary school curriculum was largely to blame, because "Many of their programmes seem to be based on the needs of kids 20 years ago. I think you'll find it is the performance of (secondary) schools, not students, that is the problem."[33]

Thus, two decades after the Picot Report and *Tomorrow's Schools*, New Zealand's state secondary schools still face similar, essentially ahistorical criticisms, to those continuously leveled at them throughout the post–World War Two era. In this respect, New Zealand is far from being alone. In the United States, a book by David Angus and Jeffrey Mirel, provocatively entitled *The Failed Promise of the American High School*, has argued that at the heart of the American conception of good high school education were the near-sacrosanct tenets of educational progressivism; namely that the curriculum should be differentiated according to individual needs, aspirations and interests, and that such differentiation should be accommodated within a single institution. This legacy, they claimed, had resulted in an ongoing mediocrity in academic attainment, with consequent disadvantages for working-class and minority students. Moreover, educational professionals, through their capture of the curriculum making process and their continuing adherence to progressivist philosophies were an impediment to attempts to monitor educational outcomes, encourage higher academic standards, and make the system more internationally competitive.[34] Not surprisingly, a number of researchers have critiqued these conclusions, particularly questioning the assumption that American secondary schools today are any worse than they were in the past.[35] The New Zealand experience post-Picot, suggests that secondary school reform, however well-meaning, is unlikely to deliver on such goals, much less silence complaints.

Rhetoric and Reality: Winning the Symbolic Victory

As this research progressed, one thing became increasingly clear—the Picot Report and *Tomorrow's Schools* were in many respects the logical, even inevitable outcome of cumulative historical processes deeply rooted in the entire postwar era. Thus taken together, the first two parts of this book illustrate just how it came to pass that New Zealand education in general and secondary education in particular were, by the mid-1980s, widely regarded as being enmeshed in a serious crisis that demanded radical solutions rather than a mere tinkering with existing structures. The early postwar expansion of comprehensive state secondary education had, from the beginning, been fraught with unresolved tensions and contradictions. These were to be glaringly exposed by the economic, social, and educational crisis of the 1970s. As far as secondary education was concerned, secondary teacher militancy, changes in newspaper reporting of educational issues, liberal critiques of the traditional secondary school curriculum, exposes of educational inequities from radical academics, and the campaigns waged by the Moral Right, all provided significant catalysts for the steady growth of national concern. In addition Māori activism, increasing support for biculturalism among educators, and calls for a separate Māori education system based on choice, lent a powerful moral and equity dimension to calls for reform based on choice. The policy environment in the decade prior to *Tomorrow's Schools* was therefore, destined to be transformed by a variety of factors. These included the Educational Development Conference, the McCombs report, the OECD report, cumulative parliamentary and political pressure for educational changes, increasing Treasury/SSC involvement in education, and the changing policy environment that gave rise to a number of reports that embodied concepts and ideas that were to be adopted by the Picot Taskforce.

Seen in their historical context, therefore, the assumptions and the recommendations of the Picot Report together with the ensuing debate analyzed in the third part of this book, turn out to be utterly unsurprising. So much so, in fact, that a relatively literate person broadly familiar with the history of postwar state secondary education in New Zealand, and furnished with a reasonable computer, might be reasonably be expected to re-create the Report, even if they had never before set eyes upon the document. The relevant concepts and phrases could be readily cut and pasted from the major reports, inquiries, commissions and advocacy group concerns from the mid-1970s through until the mid-1980s.

All this is not to overlook the major impact that neoliberal ideology was to have on state sector reform in general and educational reform in particular. Significant intellectual and personnel changes within Treasury had already taken place during the 1960s and early 1970s. It only needed the financial crisis that occurred during the second half of the 1970s to provide the context for Treasury to join the swelling ranks of educational critics, and to subsequently apply increasingly effective political pressure on the Department of Education. Beginning with the public expenditure cutbacks sought by the Muldoon government during the late 1970s and coinciding with Quigley's tenure as Associate Minister of Finance, Treasury became particularly enterprising in its determination to publicly expose the centralized education bureaucracy as both wasteful and inefficient.

This pressure became even more intensive under Labour, post-1984. Throughout the entire period the Picot Taskforce was meeting, Treasury officials literally bombarded it with papers advocating neoliberal solutions for virtually every educational problem. The direct impact of these often relatively short papers, however, is somewhat debatable. Picot himself was later to emphatically dismiss any suggestion that the taskforce had an agenda for privatizing education as "unmitigated hogwash. Never ever was it considered by our people as being a remote possibility."[36] Gianotti also supports the view that the taskforce came to its own conclusions—they "were not just ciphers of Treasury and SSC."[37]

In any case, the whole tenor of report writing during the first half of the 1980s would inevitably have impelled the taskforce toward a radical, devolved solution for education, even if Treasury had remained silent. Again, as we have seen, the taskforce appears to have been clearly resolved even by its initial meeting, that the preferred course of action was to begin with a proverbial clean slate. Hence, any subsequent overly intrusive attempt by Treasury to influence proceedings was more likely to annoy rather than to convert taskforce members—indeed there are some indications from the records of taskforce deliberations that this situation actually occurred on at least one occasion. Following the setting up of the new MoE, however, Treasury and SSC pressure was both pervasive and effective, although it was contested, sometimes bitterly so.

The main problem with what has long remained the received view in education policy research circles, therefore, is not that it singles out Treasury and SSC influence as being significant, but that it is too often claimed to be the *only* influence. To adhere to this view, I would argue, is to discount the historical and cumulative impact of so many other influences upon the reforms, to misunderstand the nature of political discourse, and to grossly oversimplify the complex process of public policymaking in liberal democratic states such as New Zealand. As we have seen, unresolved issues and tensions rendered the reforms of the 1940s that gave rise to comprehensive institutions in the first place, problematical from the very beginning. A rapidly changing social, economic, and policy environment made change inevitable in the longer-term, and the longer it was delayed, the more chance it had of being radical in prescription.

From the mid-1970s, everything seemed to be conspiring against the continuation of the status quo in secondary education. Hence, we see a wider economic and cultural crisis of the New Zealand state, coupled with largely unsuccessful attempts by embattled governments to contain public expenditure. We see a new, aggressive assertiveness on the part of Treasury and SSC that is in part a result of ideological shifts, and in part a pragmatic response to pressures imposed on them by the exigencies of the times. We also see Māori and Left-liberal suspicion of what they saw as a highly conservative, racist, and unresponsive educational structure. All this is accompanied by widespread public disenchantment, fuelled by adverse media reporting of education. Given the confluence of all these factors by the first half of the 1980s, state secondary education inevitably came to be widely regarded as being part of the problem rather than the solution.

Questions relating to the exact nature and extent of neoliberal influence on the reforms are thus difficult to answer with precision for several reasons. One of the major problems confronting any historian who begins to research the Picot Report and *Tomorrow's Schools* concerns the difficulty of separating facts from the sea of propaganda and prejudice that surrounds them. Even before the Picot Report was released vigorous campaigns were being waged through the media, aimed at galvanizing public opinion for or against the reforms. The Labour government, through the Prime Minister's Office and Logos, undertook an unprecedented public relations offensive seeking to garner public support for the reforms while attempting to minimize the impact of possible opponents. The teacher unions waged an equally ambitious campaign aimed at representing the taskforce in general and its chairman in particular as the mere tools of a shadowy all-pervasive neoliberal ideology. The success of the first may well account for the relatively muted public reaction toward the imminent demise of the country's long-standing education system following the release of *Tomorrow's Schools*, while the success of the latter is demonstrated by the fact that so many in the education sector remain convinced that both the Picot Report and *Tomorrow's Schools* are entirely neoliberal in concept, thus providing a convenient scapegoat for all the problems they face today in their professional work.

Moreover, the brief period between the release of the Report and the publication of *Tomorrow's Schools* was to give rise to further charges and counter-charges. As we have seen, many of these were politically motivated. Lange's widely reported concerns about creeping neoliberalism were symptomatic of the growing rift between himself and Roger Douglas. There is also some evidence that Lange was, even then, attempting to publicly secure a place in history which would exonerate him from any responsibility while shifting the blame for perceived problems squarely on to a shadowy and malevolent conspiracy involving Douglas, Treasury/SSC, and an imported neoliberal ideology. Similarly, some members of the Picot Taskforce appear to have used the media to draw public attention to what they feared might be Treasury and SSC interference with the original Picot recommendations, especially once the report had left their hands and been passed on to the Officials Committee and the Cabinet Equity Committee.

There are several further ironies in the struggle for control that followed the enactment of reform legislation and the subsequent change of government that brought National to power in late 1990. One of these is that Treasury and SSC attempts to influence the direction of the reforms appear to have been at its most intense at precisely this time. Within both state agencies, there was a feeling that the reforms had been somehow thwarted by an overtly Left-liberal education bureaucracy allied to the teacher's unions, determined to cling on to their old privileges. This feeling was curiously paralleled outside immediate government circles by an upsurge of critique from within education, charging that the *Tomorrow's Schools* had ushered in more rather than less centralization, leaving the bulk of the educational decision-making process to the state-controlled bureaucracy.

Even where it is possible to delve beneath charge and counter-charge, the specifically neoliberal features of the Picot report and *Tomorrow's Schools* remain difficult to isolate. The concept of provider capture and the consequent separation of policymaking from policy implementation are often claimed to be important tenets of neoliberal discourses, but they are equally part of the fabric of the Erebus Inquiry, the 1985 Defence Inquiry, the Cartwright Report, and the Fargher-Probine Report, none of which are commonly held to be "neoliberal." The justification for the sweeping changes recommended by such reports and inquiries—that it was an attempt to avoid the problem where those who supplied state services pursued their own interests at the expense of clients or consumers—was just as common to Māori, feminist, and other Left-liberal critics of secondary education as it was to Treasury and SSC officials over much the same period. Given the plethora of reports and inquiries that argued that airline officials, defense chiefs, doctors, educational bureaucrats, and even classroom teachers had actively contrived to retain their privileges while thwarting the public interest with particularly detrimental outcomes for disadvantaged groups, it was hardly surprising that the Picot Taskforce was rather quick to subscribe to a similar position that held teachers to be no longer entirely trustworthy to act as disinterested professionals. As for consumer choice, another neoliberal tenet, many morally conservative, feminist, and Māori commentators from the mid-1970s on were arguing with increasing effect for the right of parents to select appropriate schools for their children, on the dubious but then largely unquestioned grounds that the schools that did not adequately cater for their interests would automatically go under.

Such was the impact of these views that even the voucher system (which eventually proved to be a nonstarter in New Zealand), received an initially strong impetus through the success and ongoing claims of Te Kōhanga Reo, regarded within the Prime Minister's Department and elsewhere as a blueprint for any subsequent spread of the voucher concept.

Those who attempt to list the main tenets of neoliberalism often point to its emphatic denial of community and its single-minded promotion of a situation where everyone pursues their own self-interest in a free market as being a major feature. Thus, British prime minister Margaret Thatcher has been widely reported as having stated that there was no such thing as society.[38] Unlike the situation in the United Kingdom, however, much of the rationale for the Picot Report and *Tomorrow's Schools* in New Zealand seems to have been more about *re-creating* the community than denying its existence. This newfound fascination with "community" in the New Zealand of the late 1970s and early 1980s was again of decidedly mixed ideological parentage. It can be seen clearly in Franklin's rather gloomy assessment of New Zealand society, economy and culture in the late 1970s, in the academic commentaries of several prominent educational researchers, and in the papers written by Picot himself, as he desperately sought ways to reintroduce a sense of shared understandings and common interests in the midst of the nation's fraught industrial and social relations. An equally strong albeit somewhat naive fascination with the rediscovery of "community" was an integral part of Karl Polanyi's long-standing fascination with non-Western and pre-industrial tribalism; and it will be recalled that Polanyi was one of Benton's models for the justification of Māori educational autonomy in the mid-1980s. This fascination also underpinned other New Zealand initiatives during this period, such as the failed 1980s experiment with community policing, where police were expected to share the burden of fighting crime with the public.[39] And we should not forget that in New Zealand at least, the work of Foucault provided an intellectual justification, along with neoliberal concerns with financial efficiency, for deinstitutionalizing those with mental problems and placing them back in the community. Similar concerns with reactivating community involvement were epitomized in the Community Education Initiative Scheme (CEIS), which rapidly became a preferred solution for both Treasury neoliberals and Left-liberals alike.

Moreover, in New Zealand at least, isolating specifically neoliberal influences is rendered even more problematic due to the fact that educational policymaking during the 1980s manifests a remarkable continuity of personalities, many of whom espoused views that drew upon an eclectic mixture of ideologies. Sir Frank Holmes had been a strong influence on the "liberal" Currie Report but in 1978, as Chairman of the Planning

Council, he approved a Report from the Task Force on Economic and Social Planning prepared by Treasury that advocated radically revised public sector financial reforms. This association together with his long-standing friendship with Treasury's Henry Lang, provoked bitter charges from the heads of several government departments that he was in league with Treasury neoliberals. Phillida Bunkle and Sandra Coney were, in the early 1980s, widely held to be radical feminists, but their claims of provider capture by male doctors at Auckland Women's Hospital provoked a report that in its recommendations and language anticipated the neoliberal Gibbs Report. Judith Aitken, and Margaret Bazley were both not only strongly associated with women's issues during the era of second-wave feminism, but also espoused the tenets of (neoliberal?) managerialism. In the 1970s Noel Scott was expressing considerable sympathy for the plight of secondary teachers caught between the demands of the educational bureaucracy and their local communities, yet he later headed a committee on teaching that included the neoliberal National education spokesperson, Ruth Richardson. This report recommended significant parental input into professional appointments, and a system of performance appraisal, both commonly labeled "neoliberal" by many educators. Ray Fargher had wide experience and sympathy with second chance students in post-compulsory education, yet he was the joint author of a report that was to have a major influence on the Picot Taskforce. Picot himself passionately advocated conciliation and a broader sense of community as an antidote to what he saw as the destructive confrontation that characterized New Zealand during the early 1980s, yet immediately following the release of the taskforce report he was being dismissed by the teacher unions as a neoliberal stooge.

Finally, central to any understanding of neoliberalism as viewed by its critics, is the concept of hegemony. Some New Zealand education historians have argued that a major weakness of the original concept was that it tended to deny agency on those deemed to be controlled in some way, by dominant groups.[40] In response, education policy research has come to refer to "ideological coalitions," thus emphasizing the ways in which new hegemonic accords are broadened through appearing to embrace the fears and concerns of working class, new middle class and minorities.[41]

Although this concession goes some distance toward making the concept more acceptable to historians, however, it still embodies the conviction that most of those who participate in educational dialogues are limited by a kind of false consciousness that blinds them to the socioeconomic realities that critical theorists alone can identify. In the immediate run-up to the creation of the Picot Taskforce and again during its controversial aftermath, however, there is ample evidence of a much more active agency than even this extended definition of hegemony allows. The main impression one gathers from the large number of public submissions to the Picot report for instance, is that most people seemed to be shrewdly aware of just what was going on. Far from being taken in by the rhetoric, they frequently employed quite sophisticated skills of negotiation in order to wring the best deal they could from the proposed new structure and its underlying assumptions.

Governments too were active agents rather than mere Treasury pawns in this period, even in the face of apparently irresistible global trends such as neoliberalism. Michael Mintrom and John Wanna have recently observed that both New Zealand and Australia were at the at forefront at adapting key components in their structures of governance to the new world order with the intention of making their economies more internationally competitive. They began with deregulation of the financial sector and removing tariffs, but went on to pioneer managerialism and new public sector management during the later 1980s and 1990s, intended to increase flexibility and responsiveness, orientate public services to a results culture, and to heighten political control and management of state agencies.[42] Mintrom and Wanna concluded, however, that Antipodean governments "have not resigned themselves to the buffeting forces of globalization, but rather turned these to national advantage." Hence, recent governmental action had been "less associated with ideology or party political complexion and more with new thinking about preferred government arrangements."[43]

One fruitful approach to this vexed problem of agency may well be for future research to examine more closely the actual discourses that have become associated with the reforms, using critical discourse analysis (CDA) in new, historically informed ways. The notion that language can be used for self-interested ends by various power groups in society originally stemmed from the Frankfort School and particularly from critical theorist such as Habermas. From Foucault and others, came the concept of language as a form of social action.[44] One of the difficulties for historians seeking to understand the intricacies of public policy rhetoric in New Zealand, however, is that although CDA has often been used to critically interrogate neoliberal and racist discourses, it has rarely been employed to examine the way in which discourses from across apparently oppositional ideologies can intersect at decisive moments in time.

Norman Fairclough is one recent scholar who has pointed to the uses of CDA in transdisciplinary research. Fairclough claims that "through focusing on discourse CDA aims to elucidate the discoursal moment of social processes, practices, and change in its dialectical relations with other moments."[45] It has been pointed out, however, that the form and content of discourses differs substantially, depending on the commentator's appreciation of the problem-situation.[46] The period leading up to the New Zealand educational reforms of the late 1980s demonstrates that virtually all the key groupings of critics had a high degree of certainty about the problem and how to deal with it. In turn this reflected a high level of system knowledge, which Renwick himself acknowledged when he referred to the power of the various education lobbies in the early 1980s. What seems evident in New Zealand is that, beginning in the second half of the 1970s and continuing more emphatically through the 1980s and beyond, political entrepreneurs on both sides of the ideological divide discovered at much the same time, a common discourse—a language of power that enabled them to break the bureaucratic stranglehold on policymaking; in the process catapulting themselves to the political center-stage.

Of particular interest here are the interactions between those groupings that, at first glance, appear to be unlikely allies. H.M. Kliebard, in his timely afterword to his third edition of *The Struggle for the American Curriculum*,[47] has recently outlined a set of issues broadly familiar to those facing historians researching the Picot era. Citing sources such as Filene, Kliebard identified "shifting coalitions around particular issues" whose stand was often prompted by opportunism and improvisation rather than a consistent ideology.[48] Drawing upon the work of Rodgers and others allowed him to take the concept a stage further by arguing that reformers were held together by a common faith in their ability to use particular languages in order to build a constituency.[49] This both permitted them to identify each other as being committed to a similar reformist cause and to create a temporary coalition of otherwise diverse subgroups behind a particular reform, thereby producing a community of discourse.[50]

Viewed in this way, the various groupings that critiqued New Zealand secondary education from the mid-1970s through the 1980s can be seen, not simply in terms of what divided them, but rather by what united them; specifically how those holding opposing ideological positions came together at a crucial juncture to support educational reform—even though they saw different things in it for themselves and those they claimed to speak for.[51] Furthermore, in a relatively small New Zealand society and within a relatively compact Wellington-based policy environment, where personal relationships often cut across political differences, both Left and Right were able to pluck ideas from a common policy discourse that was rapidly becoming dominant in the first half of the 1980s to the extent that would have been vastly more difficult in the mass-societies of the Northern Hemisphere.

All this raises some key issues for future research. Colin James is one political commentator who argues that the election of Labour in 1984 brought the so-called Vietnam generation to political prominence. In his view these were "big picture" people who actively sought the liberalization of what they regarded as an overly restrictive society.[52] To take one relevant example here—calls for liberalization of existing laws against abortion and homosexuality prior to the election of Labour in 1984 primarily focused on individual choice and human rights, posing some difficult questions for historical research into the reform period. To what extent were advocates of choice in these areas prepared to compromise with those who embraced free choice and individual responsibility in activities as diverse as finance, health and education? And when one turns to key movers and shakers of the era, just how easy was it to move from say, social activism, to neoliberalism, however we choose to define these?

A key factor in marshalling pressure for educational reform, and one that seems more applicable to small democracies such as New Zealand than larger societies such as the United Kingdom or the United States, was the very direct relationship between the politicians, the media, and the electorate as a whole that was a feature of the 1980s. At about this time, deregulation of the mass media brought forth a plethora of niche magazines catering for a politically interested reading public. In turn this seems to have served to sharpen the impact of the social, economic, and cultural crisis that in turn provided the backdrop for the dissemination of a common educational critique. The exigencies of the times and the perceived need to impose some sense of order on the apparently chaotic state of secondary education in particular underlines the fact that there were several competing groups in New Zealand during 1980s, each of which shared an ambiguous attitude toward the state as constituting not only the source of the problems they identified, but also the means by which they sought redress. Not surprisingly, no single grouping succeeded in getting all that they desired from the reforms. In any case purist aims in education have been notoriously unachievable in New Zealand, but the situation was hardly helped by the fact that so many demands on secondary education were both ambivalent and contradictory. Ideologically opposed critics may have utilized a common discourse in order to encourage the schools to emphasize clear goals in curricula and in pedagogy, for demonstrable outcomes. But they were at the same time often deeply suspicious of centralization, each suspecting the other of attempting to capture the system for their own ends.

Ultimately, however, it is at least arguable that the motives behind the reforms were less about concrete and measurable outcomes than they were about symbolic victory. It is perhaps difficult to avoid the conclusion that, in New Zealand, the whole the debate over reform was ultimately about whose values and beliefs would achieve the legitimation and respect that acceptance into a national discourse inevitably provides. Kliebard's conclusion about the various interest groups who contested the American curriculum may well be true of those who so bitterly contested, and continue to contest, the Picot reforms and their outcome, because they too were part of

> a national morality play in which those hopes and fears were enacted. The articulation of those platforms thereby precipitated a searing conflict which ultimately was fought over whose deep-seated convictions would predominate in the emerging new society.[53]

Perhaps the true legacy of Picot and *Tomorrow's Schools* then, is that they continue to serve as both focal point and catalyst for the long-standing struggle about how we will in future view the world, education, and secondary schooling. Certainly, from an early twenty-first century perspective, the result seems to be an untidy and not wholly satisfactory compromise. Even a cursory glance at contemporary educational structures, policy documents and curriculum statements reveals a bewildering muddle of neoliberal, bicultural, constructivist and equity concepts, with little or no acknowledgement of any contradictions between them.

Despite continuing criticism of the Picot Report and *Tomorrow's Schools*, however, it is highly unlikely that New Zealand will move back to anything approximating the old three-tiered education system. Examining the conditions under which any new paradigm is likely to be successfully challenged by a fresh policy discourse, Burns and Carson argue that when the original reforms are seen as failing to address the problems they originally claimed to be able to remedy, those who have not "invested" in the new paradigm are more likely to be open to pressure and persuasion from critics.[54] In the case of *Tomorrow's Schools*, there would seem to be too much investment in the new educational paradigm from across the political spectrum for this to happen any time soon. Hence, like the 1877 Education Act, the reforms are likely to be increasingly cemented in place, although some form of tinkering with existing arrangements cannot be ruled out. The bulk funding issue remains dormant but is probably far from being dead, especially given the importance a previous National Government placed on it. There has also been some talk of making further provision for struggling schools to pool their administrative resources, as in a single "super-BoT," for instance. And as with its late-nineteenth-century counterpart, the new reforms have left education in general and secondary education in particular with unanswered tensions and questions about future funding arrangements, public versus private schools, the balance between private and public good, between equity and efficiency, and between state priorities and local autonomy for all New Zealanders in the new century.

This latter tension is arguably the most significant legacy of the reforms. As Leicester Webb so astutely put it some seventy years ago, the basic problem of education is, broadly speaking,

> the problem of using the administrative resources and the coordinating power of the central government and at the same time curbing the central government's inherent love of uniformities. It is the problem of admitting the nation state to its preponderating share in the control of education and at the same time keeping at a distance those who, from time to time, identify themselves with the state and profess to act in its name. It is the problem, perhaps, of persuading the nation state to support education without requiring education to support it slavishly in return. An education system should strike its roots downwards into the social life of which the nation is composed. But it should also reach upwards towards a social order which is only adumbrated in things as they are.[55]

Equally perceptively, Webb saw in New Zealand "a general and unfortunate willingness" to seek solutions for educational problems in terms of institutions alone. For him no scheme of educational reform would be worth anything if we lost sight of that essential truth that no administrative system could make up for shortcomings elsewhere. Thus, at the end of this book, we come around full circle to the question posed by the rival positions on the reforms. Who really won? When the Butterworth's posed the question as to whether the reforms were or were not, "a famous victory," they wisely avoided a definitive answer. Rosslyn Noonan's verdict on the confrontation between what she labeled "participatory democracy" and "New Right ideology" that immediately followed the release of the Picot Report it was that it all ended, she said, in a draw.[56]

Such equivocation in naming a victor in these struggles between Right and Left is perhaps justifiable. Indeed on the basis of the evidence presented in this book, the whole question of who really won seems largely irrelevant. Edwards and Moore have recently concluded that while some academics have postulated an essential incongruity between the implementation of the right-wing economic policies and socially liberal reforms during the decisive period of New Zealand's fourth Labour government, it is actually no accident that the two trends occurred at the same time.[57] As we have seen, beginning in the 1970s and continuing through the 1980s, neoliberal and Left-liberal pressure for change drew upon a commonly created discourse that not only profoundly influenced social policy at the time but also went on to give birth to a new blended ideology that continues to shape the way we view the world. Herein lies the major reason for revisiting New Zealand's educational reforms in the context of secondary schooling and public policy.

Notes

1 To Suit a Political Purpose? Reinterpreting the Educational Reforms

1. G. Butterworth and S. Butterworth, *Reforming Education. The New Zealand Experience 1984–1996*, Palmerston North: Dunmore Press, 1998, Chapter 4.
2. Report of the Taskforce to Review Educational Administration, *Administering for Excellence. Effective Administration in Education*, Wellington: Government Printer, 1988, 22. Henceforward cited as the Picot Report.
3. Ibid., 81.
4. Ibid., 82–83.
5. F. Holmes, "Lay and Professional Participation in Educational Administration," unpublished paper, c. early 1976, BCDQ, 5193R, A739, Archives New Zealand, Auckland.
6. S. Smelt, *Today's Schools. Governance and Quality*, Wellington: Institute of Policy Studies, Victoria University, 1988, 74.
7. Ibid., 74.
8. Ibid.
9. R. Kerr, "Stand and Deliver. A Business leader Challenges the Education Establishment to Enter the Real World," *Metro*, 8, no. 129, March 1992, 110–115.
10. Ibid., 112–114.
11. A. Peachey, *What's Up with Our Schools? A New Zealand Principal Speaks Out*, Glenfield, Auckland: Random House, 2005, 84.
12. Ibid., 89.
13. T. Watkin, "Class Action," *New Zealand Listener*, October 11, 1997, 31.
14. H. Lauder, S. Middleton, J. Boston, and C. Wylie, "The Third Wave; a Critique of the New Zealand Treasury's Report on Education," *New Zealand Journal of Educational Studies*, 23, no. 1, 1988, 15–43; J. Boston, B. Haig, and H. Lauder, "The Third Wave: A Critique of the New Zealand Treasury's Report on Education, Part 11," *New Zealand Journal of Educational Studies*, 23, no. 2, 1988, 115–144.
15. L. Gordon, and J. Codd, *Education Policy and the Changing Role of the State*, Proceedings of the New Zealand Association for Research in Education

Seminar on Education Policy, Massey University, July 1990, Delta Studies in Education, no. 1, Massey University: Palmerston North, 1990.
16. Lauder et al., 15.
17. See, for instance, G. McCulloch, "Serpent in the Garden: Conservative Protest, the 'New Right' and New Zealand Educational History," *History of Education Review*, 20, no. 1, 1991, 73–97.
18. I. Snook, "Educational Reform in New Zealand: What *Is* Going On?" *Access*, 8, no. 2, 1989, 9–18; Ibid., 75.
19. G. Grace, "Labour and Education: The Crisis and Settlements of Education Policy," in *The Fourth Labour Government. Radical Policies in New Zealand*, second edition, edited by M. Holland and J. Boston, Auckland: Oxford University Press, 1990, 165–191; G. Grace, "The New Zealand Treasury and the Commodification of Education," in *New Zealand Education Policy Today: Critical Perspectives*, edited by S. Middleton, J. Codd, and A. Jones, Wellington: Allen and Unwin, 1990, 27–39.
20. E.A. Gordon, "Picot and the Disempowerment of Teachers," *Delta*, 41, May 1989, 23–32.
21. J. Codd, R. Harker, and R. Nash, "Introduction: Education, Politics and the Economic Crisis," in *Political Issues in New Zealand Education*, edited by J. Codd, R. Harker, and R. Nash, Palmerston North: Education Palmerston North, 1990, 7–22.
22. J. Codd and L. Gordon, "School Charters: The Contractualist State and Education Policy." *New Zealand Journal of Education Studies*, 26, no. 1, 1991, 21.
23. R. Dale, "Regulation Theory, Settlement and Circulation Policy" in *Education Policy and the Changing Role of the State*, Proceedings of the New Zealand Assessment for Research on Education Seminar on Educational Policy, edited by L. Gordon and J. Codd, *Delta Studies in Circulation*, no. 1, Palmerston North: Massey University, July 1990, 33–44.
24. M. Apple, "Ideology, Equality and the New Right," in *Education Policy and the Changing Role of the State*, Proceedings of the New Zealand Association for Research on Education Seminar on Educational Policy, edited by L. Gordon and J. Codd, *Delta Studies in Education*, no. 1, Palmerston North: Massey University, July 1990, 8.
25. A.-M. O'Neill, *Curriculum Reform: The Political Context, Delta*, 48, no. 1; A.-M. O'Neill, *Curriculum Reform: Development Issues, Delta*, 48, no. 2, 1996, pp. 1–9.
26. M. Thrupp, Introduction to *New Zealand Journal of Educational Studies*, Special Issue: A Decade of Reform in New Zealand Education: Where to Now? 34, no. 1, 1999, 5–7.
27. P. Fitzsimons, M. Peters, and P. Roberts, *New Zealand Journal of Educational Studies*, 34, no. 1, 1999, 35, 42.
28. J. Codd, "Educational Reform and the Culture of Distrust," *New Zealand Journal of Educational Studies*, 34, no. 1, 1999, 48, 52.
29. J. Boston (ed.), *The State under Contract*. Wellington: Bridget Williams Books, 1995, Preface, ix.

30. Ibid., x.
31. M. Olssen, "Restructuring New Zealand Education: Insights from the Work of Ruth Jonathan," *New Zealand Journal of Educational Studies*, 34, no. 1, 1999, 63.
32. M. Olssen, J. Codd, and A.-M. O'Neill, *Education Policy: Globalization, Citizenship and Democracy*, London: Sage, 2003.
33. Olssen et al., 13.
34. Ibid., 15.
35. J. Gilbert, *Catching the Knowledge Wave? The Knowledge Society and the Future of Education*, Wellington: NZCER Press, 2005.
36. For a critique of the Ministry's adoption of this philosophy, see J. Clark, "The Strange Case of the Ministry of Education's Mysterious Philosophy of the Curriculum," *Delta*, 51, no. 1, 2000, 41–53.
37. Butterworth and Butterworth, *Reforming Education*, 233, 254.
38. Ibid., 10, 253.
39. D. McKenzie, "Final Fulfillment in Managerial Nirvana. A Review of *Reforming Education: The New Zealand Experience, 1984–1996*," *New Zealand Education Review*, March 26, 1999, 10.
40. G. Butterworth and S. Butterworth, "Bad Review Reply," *New Zealand Education Review*, July 30, 1999, 8.
41. G. Lee and H. Lee, "Education Policy Dynamics Ignored," *New Zealand Education Review*, August 27, 1999, 8.
42. W. Richardson, "Historians and Educationalists: The History of Education as a Field of Study in Post-war England, Part 11: 1972–96," *History of Education*, 28, no. 2, 1999, 130–131.
43. See G.R. Elton, "Presidential Address: The Historian's Social Function," *Transactions of the Royal Historical Society*, 5th series, 27, 1977, 208, 210, 204.
44. Richardson, 122. See also B. Simon, "Education and Social Change: A Marxist Perspective," *Marxism Today*, February 1977, 37–43.
45. H. Silver, *Education as History: Interpreting Nineteenth and Twentieth Century Education*, London & New York: Methuen, 1983.
46. P. Goldsmith, "Review of *Treasury: The New Zealand Treasury, 1840–2000* by Malcolm McKinnon, Auckland University Press: Auckland, 2003," *New Zealand Journal of History*, 38, no. 1, 2004, 82–83.
47. D. Thomson, *Selfish Generations? The Ageing of New Zealand's Welfare State*, Wellington: Bridget Williams Books, Introduction, x.
48. C. James, *New Territory. The Transformation of New Zealand, 1984–92*, Wellington: Bridget Williams Books, 1992, 80.
49. Ibid., 79.
50. I. Snook, "Policy Change in Higher Education. The New Zealand Experience," *Higher Education*, 21, 1991, 625.
51. McCulloch "Serpent in the Garden."
52. J. Barrington, "Historical Factors for Change in Education," in *Redistribution of Power? Devolution in New Zealand*, edited by P. McLinlay, Wellington: Victoria University Press for Institute of Policy Studies, 1990, 191.
53. McCulloch, "Serpent in the Garden," 73–97.

54. R. Openshaw, *Unresolved Struggle. Consensus and Conflict in State Post-primary Education*, Palmerston North: Dunmore Press, 1995.
55. G. McCulloch, *Documentary Research in Education, History and the Social Sciences*, London: Routledge Falmer, 2004, 6.
56. R. Openshaw, "Preparing for Picot: Revisiting the 'Neoliberal' Education Reforms," *New Zealand Journal of Educational Studies*, 38, no. 2, 2003, 135–150.
57. See, for instance, D. Roulston, "Educational Policy Change, Newspapers and Public Opinion in New Zealand, 1988–1999," *New Zealand Journal of Educational Studies*, 41, no. 2, 2006, 145–162. See also D. Roulston, "An Historical Study of Newspapers and Public Opinion about Education," MA diss., Wellington: Victoria University, 1986.
58. K. Jenkins, *Rethinking History*, London and New York: Routledge, 1991, 70
59. R. Evans, *Defence of History*, London: Granta Books, 1997, 219.
60. R. Dawkins, *The God Delusion*, London: Bantam Press, 2006, 361.

2 Almost Alone in the World, 1942–1968

1. See, for instance, R. Openshaw, *Unresolved Struggle. Consensus and Conflict in State Post-primary Education*, Palmerston North: Dunmore Press, 1995.
2. G. Grace, "Labour and Education: The Crisis and Settlements of Education Policy," in *The Fourth Labour Government. Radical Politics in New Zealand*, edited by M. Holland and J. Boston, second edition, Auckland: Oxford University Press, 165.
3. G.D. Lee, "From Rhetoric to Reality: A History of the Development of the Common Core Curriculum in New Zealand Post-Primary Schools," PhD diss., University of Otago, 1991; D. McKenzie, H. Lee, and G. Lee, *Scholars or Dollars? Selected Historical Case Studies of Opportunity Costs in New Zealand Education*, Palmerston North: Dunmore Press, 1996; R. Openshaw, G. Lee, and H. Lee, *Challenging the Myths. Rethinking New Zealand's Educational History*, Palmerston North: Dunmore Press, 1993.
4. R. Openshaw, G. Lee, and H. Lee, "The Comprehensive Ideal in New Zealand Education: Challenges and Prospects," in *Death of the Comprehensive High School: Historical, Contemporary and Comparatives Perspectives*, edited by B.M. Franklin and G. McCulloch, New York and London: Palgrave McMillan, 2008.
5. G.N. Marshall, "The Development of Secondary Education in New Zealand from 1935 to 1970," PhD diss., Hamilton: University of Waikato, 1989, 121.
6. See particularly, Openshaw, *Unresolved Struggle*, 48.
7. *Appendices to the Journal of the House of Representatives*, 1946, E-1, 9. Hereafter cited as *AJHR*.

8. G. Dunstall, "The Social Pattern," in *The Oxford History of New Zealand*, edited by G.W. Rice, second edition, Auckland: Oxford University Press, 1991, 452.
9. Ibid.; G. Hawke, "Economic Trends and Economic Policy, 1938–1992," in *The Oxford History of New Zealand*, edited by R.W. Rice, second edition, Auckland: Oxford University Press, 1992, 432.
10. Dunstall, "The Social Pattern," 452.
11. C. James, *New Territory. The Transformation of New Zealand, 1984–92*, Wellington: Bridget Williams, 1992, 19.
12. Dunstall, "The Social Pattern," 465.
13. See, for instance, Centre for Contemporary Studies, *Unpopular Education. Schooling and Social Democracy in England since 1945*, London: Hutchinson, 1981, 144–145.
14. Lee, "From Rhetoric to Reality," 498–499.
15. Openshaw, *Unresolved Struggle*, 52.
16. See P. Miller, *Long Division. State Schooling in South Australian Society*, Netley, South Australia: Wakefield Press, 1986, 208–230. See also Centre for Contemporary Studies, *Unpopular Education*.
17. Lee, "From Rhetoric to Reality," 499.
18. *AJHR*, 1943, E-1, 2–3, cited ibid., 499–500.
19. *Dominion*, March 15, 1943, editorial. Cited in Openshaw, *Unresolved Struggle*, 54.
20. Thomas Report, 5.
21. I.A. McLaren, *Education in a Small Democracy: New Zealand*, World Education Series, London: Routledge and Kegan Paul, 1974, 128.
22. Lee, Lee, and Openshaw, "The Comprehensive Ideal," 169–184.
23. C. Andrews, "An Analysis of Some of the Major Changes in Girls' Secondary Education since World War Two," MA diss. University of Auckland, 1980, 25; See also S. Middleton, "Feminism and Education in Post-war New Zealand," in *Reinterpreting the Educational Past*, edited by R. Openshaw and D. McKenzie, Wellington: NZCER, 1987, 132–146.
24. R. Openshaw and M. Walshaw, "Pandering to Women? The Introduction of the Science and Mathematics New Scheme in an Era of Ambiguity," *New Zealand Journal of Educational Studies*, 41, no. 2, 2006, 241–255.
25. R. Openshaw, *Schooling in the '40s and '50s: An Oral History*, Research Resources No. 1, Palmerston North, Educational Research and Development Centre, 1991, 17.
26. B. Cocklin, "Separate or Mixed Schooling: A Revisionist Study of Secondary Education in Marlborough 1946-58," MA diss. Palmerston North: Massey University, 1983, 6.
27. Middleton, "Feminism and Education in Post-war New Zealand," 513.
28. J. Barrington, "'Learning the Dignity of Labour': Secondary Education Policy for Māoris," *New Zealand Journal of Educational Studies*, 23, no. 1, 1988, 55.
29. Ibid., 56.

30. Ibid., 54.
31. Cited in Openshaw, *Unresolved Struggle*, 71–72.
32. Ibid.
33. Ibid.
34. Ibid., 72–73.
35. Ibid.
36. E.A. Gordon, "Access. The Limits and Capacity of the State," DPhil. diss., Palmerston North: Massey University, 1989, 110.
37. Lee, "From Rhetoric to Reality," 547–548.
38. Ibid., 549–550.
39. Ibid., 547–548.
40. Ibid., 513.
41. Ibid., 514.
42. C.E. Beeby, *The Biography of an Idea. Beeby on Education*, Educational Research Series No. 69, Wellington: NZCER, 1992, 10.
43. W.B. Sutch, *The Quest for Security in New Zealand*, London: Oxford University Press, 1966, 27.
44. C. Whitehead, "The Thomas Report: A Study in Educational Reform," *New Zealand Journal of Educational Studies*, 9, no. 1, 1974, 162.
45. R. Shuker, The *One Best System? A Revisionist History of State Schooling in New Zealand*, Palmerston North: Dunmore Press, 1987, 163.
46. Marshall, "Development of Secondary Education," 241–242.
47. Ibid., 132–133, 243–244.
48. Cited in Openshaw, *Schooling in the '40s and '50s*, 18.
49. C.E. Beeby, "New Zealand—An Example of Secondary Education Without Selection," *International Review of Education*, 2, no. 4, 1956, 396, cited by G. McCulloch, "Secondary Education without Selection? School Zoning Policy in Auckland since 1945," *New Zealand Journal of Educational Studies*, 21, no. 2, November 1986, 96.
50. McCulloch, "Secondary Education without Selection," 100.
51. Ibid., 102–103.
52. Ibid.
53. *PPTA Journal*, 1, no. 8, November 1955, editorial.
54. Openshaw, Lee, and Lee, *Challenging the Myths*, 215–216. The term "new" is used here to distinguish this post-war qualification from the short-lived "old" School Certificate instituted in 1934.
55. Ibid., 217.
56. Ibid., 221–222.
57. Ibid., 230.
58. H.F. Lee, "The Credentialed Society: A History of New Zealand Public School Examinations 1871–1990," PhD diss., University of Otago, 1991, 613–614. Thus by 1945 only six out of ninty-seven district high schools had been placed on the accrediting list.
59. Ibid., 615.
60. Ibid., 618.
61. Ibid., 641.

62. Report of the Minister of Education (T.H. McCombs), 1947, *AJHR* (1948), E-1, 5.
63. C.E. Beeby, "A Special Topic. The Post-primary Curriculum," *AJHR*, 2, 1958, E-1, 5.
64. Report of the Minister of Education (T.H. McCombs), 1947, 4.
65. Ibid., 2–3.
66. Ibid., 4.
67. Ibid.
68. J.H. Murdoch, *The High Schools of New Zealand. A Critical Survey.* Christchurch: NZCER, 1944, 369, 386.
69. W.L. Renwick to C. Whitehead, April 2, 1973, 2 (3 pp.), ABEP, W4262, Box 1704, 34/1/23, Post primary Curriculum (Thomas Report).
70. *New Zealand Education Gazette*, 23, no. 9, 1994, 219. Cited in J. Soler, "Reading the Word and the World: The Politics of the New Zealand Primary School Literary Curriculum from the 1920s to the 1950s," Ph.D. diss., University of Otago, 1996, 125.
71. See, for instance, Beeby, *Biography of an Idea*, Chapter 8.
72. E.M. Blaiklock, *Between the Morning and the Afternoon*, Palmerston North: Dunmore Press, 1980, 8.
73. E.M. Blaiklock, "Whistling Out of Fashion," *New Zealand Herald*, June 27, 1978. Cited in D. More (ed.), *The Best of Grammaticus. Writings of Professor E.M. Blaiklock*, Auckland: Wilson and Horton, 1984, 54.
74. W. Anderson, *The Flight from Reason in New Zealand Education*, Auckland: Catholic Teachers' Association of Auckland, 1944, 3, 14.
75. C.J. Wood, "Education as a Preparation for Employment," *New Zealand Commerce*, 3, no. 4, October 15, 1947, 24.
76. Ibid., 27.
77. Ibid., 27.
78. A. Hanna, "Teachers Answer Back," *New Zealand Commerce*, 3, no. 4, October 15, 1947, 25.
79. See Centre for Contemporary Studies, *Unpopular Education*, 70–89.
80. B.F.R. Down, "Re-reading the History of Western Australian State Secondary Schooling after 1945," PhD diss., Murdoch University., 1993, ix–x.
81. Openshaw, *Unresolved Struggle*, 66–68.
82. Algie, *NZPD*, vol. 291, 1950, 2529.
83. J.D.S. McKenzie, G. Lee, and H. Lee, *The Transformation of the New Zealand Technical High School*, Palmerston North: Delta Research Monograph, 1990, 43.
84. N. Alcorn, *"To the Fullest Extent of His Powers," C.E. Beeby's Life in Education*, Wellington: Victoria University Press, 1999, 187–188.
85. Samuel S. Green. "The Case for a Commission of Inquiry or Consultative Committee to Investigate the Administration of Education in New Zealand," Unpublished paper (November 18, 1954). BAAA (Department of Education, Northern Regional Office, Acc A841/30. Archives New Zealand, Auckland).
86. Draft of letter sent *to New Zealand Herald*, June 19, 1956, ABEP, W4262, Box 692, Criticism, Part 1.

87. L.F. le Ensor to Beeby, June 28, 1956, ABEP, W4262, Box 692, Criticism, Part 1.
88. See, for instance, R. Openshaw, "Diverting the Flak: The Response of the New Zealand Department of Education to Curriculum Controversy," *Change. Transformations in Education*, 4, no. 1, May 2001, 33–47.
89. A.-M. O'Neill, "Reading Educational Policy," MEd diss., Palmerston North: Massey University, 1993, 149.
90. Openshaw, *Unresolved Struggle*, Chapter 10.
91. *Report of the Committee on New Zealand Universities* (The Parry Report), Wellington: Department of Education, 1959, 5.
92. A.M. Endres, "The Political Economy of W.B. Sutch: Toward a Critical Appreciation," *New Zealand Economic Papers*, 20, no. 1, 1986, 17–19; 25.
93. Parry Report, 14.
94. Ibid., 17
95. See ibid., especially Chapter 2—"The Universities and New Zealand's Economic Development"; and Chapter 3—"The Demand for and the Supply of University Graduates."
96. Ibid., 12.
97. "The Parry Report," *PPTA Journal*, 6, no. 2, March 1959, editorial, 1.
98. McLaren, *Education in a Small Democracy*, 44.
99. Hawke, "Economic Trends and Economic Policy," 432.
100. See, for instance, Table 1: National Income and Expenditure in Education, 1939–1960, *Report of the Commission on Education in New Zealand* (The Currie Report), Wellington: Government Printer, 1962, 135.
101. R. Shuker and R. Openshaw, *Youth, Media and Moral Panic in New Zealand*, Delta Research Monograph No. 11, Palmerston North: Massey University, 1990, 31.
102. Ibid., 33.
103. Ibid., 39.
104. Cited by McLaren *Education in a Small Democracy*, 1974, 120.
105. See, for instance, R. Openshaw, "Subject Construction and Reconstruction; Social Studies and the New Mathematics," in *The School Curriculum in New Zealand: History, Policy and Practice*, edited by G. McCulloch, Palmerston North: Dunmore Press, 1992, 201–208.
106. P. Meikle, *School and Nation. Post Primary Education since the War*, Wellington: New Zealand Council for Educational Research, 1961, 6–11.
107. Ibid., 18–20.
108. Ibid., 21.
109. Ibid., 38.
110. *Report of the Commission on Education in New Zealand* (The Currie Report), Wellington: Government Printer, 1962, 3.
111. D.J. Scott, "The Currie Commission and Report on Education in New Zealand, 1960–1962," PhD diss. University of Auckland, 1996.
112. Currie Report, 11.
113. The Associated Chambers of Commerce of New Zealand, Submission, 8, E.50/13, Archives New Zealand, Wellington.

114. New Zealand Engineering and Related Industries Association and Metal Trades Employers Association of Wellington (Inc.), Submissions, E50/14, Archives New Zealand, Wellington.
115. New Zealand Department of Education, Submission, No. 11A. E50/8, Archives New Zealand, Wellington.
116. Ibid.
117. G. McCulloch, "Historical Perspectives on New Schooling," in *Myths and Realities. Schooling in New Zealand*, edited by A. Jones, G. McCulloch, J. Marshall, H.S. Smith, and L.T. Smith, Palmerston North: Dunmore Press, 1990, 41.
118. J.H. Murdoch, E.50/12 Submission, Archives New Zealand, Wellington.
119. W.B. Sutch, E.50/15, Submission, Archives New Zealand, Wellington.
120. Currie Report, 12.
121. Ibid.
122. Centre for Contemporary Studies, *Unpopular Education* (Ibid), 143.
123. Ibid.
124. Currie Report, 26.
125. Ibid., 7.
126. "First Things First," *PPTA Journal*, 9, no. 8, 1962, editorial.
127. McCulloch "Historical Perspectives on New Schooling," 1990, 40.
128. Centre for Contemporary Studies, *Unpopular Education*, 121.
129. R. Fry, *It's Different for Daughters. A History of the Curriculum for Girls in New Zealand Schools, 1900–1975*, Wellington: NZCER, 1985, 186.
130. Currie Report, 394.
131. O'Neill, "Reading Education Policy," 152.
132. R. Openshaw and M. Walshaw, "Pandering to Women," 241–256.
133. Dunstall, "The Social Pattern," 478 (451–480).
134. R. Walker, "Māori People since 1950," in *The Oxford History of New Zealand*, edited by G.W. Rice, second edition, Auckland: Oxford University Press, 1992, 503 (498–519).
135. *Report on Department of Māori Affairs* (The Hunn Report), Wellington: Government Printer, 1960, Paragraph 33.
136. L.M. Kenworthy, T.B. Martindale, and S.M. Sadaraka, *Some Aspects of the Hunn Report. A Measure of Progress*, Wellington: Victoria University, School of Political Science and Public Administration, 1968, 20.
137. Māori Synod of The Presbyterian Church, *A Māori View of the Hunn Report*, Christchurch and Whakatane: Presbyterian Bookroom and Te Waka Karaitiana Press, 1961, 10.
138. Walker, "Māori People since 1950," 503.
139. G.H. Smith and L.T. Smith, "Ki Te Whai Ao, Ki Te Ao Marama: Crisis and Change in Māori Education," in *Myths and Realities. Schooling in New Zealand*, edited by A. Jones, G. McCulloch, J. Marshall, H.S. Smith, and L.T. Smith, Palmerston North: Dunmore Press, 1990, 139.
140. Currie Report (Ibid.), 401.
141. Ibid., 415.
142. Ibid., 416.

143. Currie Report (Ibid.), 418.
144. Ibid.
145. Ibid.
146. Ibid., 422.
147. Ibid., 423.
148. Ibid., 428–429.
149. Ibid., 426.
150. W.B. Sutch, *The Māori Contribution—Yesterday, Today and Tomorrow*, Wellington: Department of Industries and Commerce, 1964.
151. Sutch, *The Māori Contribution*, 39.
152. Centre for Contemporary Studies, *Unpopular Education*, 169, 190.
153. M. McKinnon, *Treasury. The New Zealand Treasury 1840–2000*, Auckland: Auckland University Press, 2003, 230–232.
154. Ibid., 227, 239.
155. Ibid., 237.
156. R.D. Muldoon, "The Future of University Education in New Zealand," *Delta*, 2, May 1968, 5. This article also includes an appended critical commentary by *Delta* editor, R.J. Bates.
157. P.J. Mundy, "The Current Crisis in New Zealand Education," *Delta*, 2, May 1968, 31.
158. "Critical Commentary," appended to Muldoon article ("Future of University Education," 7).

3 Game War

1. M. McKinnon. *Treasury. The New Zealand Treasury 1840–2000*, Auckland: Auckland University Press, 2003, 282.
2. Ibid., 283.
3. D.C. Pearson and D.C. Thorns, *Eclipse of Equality: Social Stratification in New Zealand*, Sydney: Allen and Unwin, 1983, 54–55.
4. Ibid., 81–82.
5. B. Easton, "The Resructuring and Liberalisation of the New Zealand Economy," *British Review of New Zealand Studies*, 6, November 1993, 75.
6. Ibid., 76–77, especially Figure 1.
7. S. Birks and S. Chatterjee, *The New Zealand Economy. Issues and Policies*, second edition, Palmerston North: Dunmore Press, 1992, 236.
8. *The World Atlas 1994*, Washington: World Bank, December 1994, 18–19.
9. McKinnon, *Treasury*, 284.
10. S. Harvey Franklin, *Trade, Growth and Anxiety. New Zealand beyond the Welfare State*, Wellington: Methuen, 1978.
11. Ibid., 2.
12. Ibid., 133.
13. Ibid., 94; 379–383.

14. Ibid., 383–397.
15. B. Picot, *Working Together*, Wellington: NZ Planning Council, 1978 (June), NZPC No. 7.
16. See, for instance, Brian Picot. *South Africa's Apartheid. An Attempt at Perspective*. Auckland: Epsom, November, 1980. In this privately produced paper, Picot supported the recently concluded Gleneagles Agreement, arguing that the planned tour of New Zealand by the South African rugby team should be called off.
17. Picot, *Working Together*, 4.
18. Ibid., 7.
19. Ibid., 9.
20. B. Picot, *New Zealand Attitudes to Challenge and Change. A Discussion Paper*. Auckland: Brian Picot, March 1979.
21. Ibid., 3.
22. Ibid., 27.
23. J. Barrington, "The Politics of School. Government," in *The Politics of Education in New Zealand*, edited by M. Clark, Wellington: NZCER, 66–68.
24. D. Roulston, "An Historical Study of Newspapers and Public Opinion about Education," MA diss., Wellington: Victoria University, 1986.
25. Ibid., 26, 51, 86.
26. Staff Reporter, "Education Spending: Is Record Enough?" *Christchurch Star*, August 23, 1977, 1.
27. "Kids Back in Class but Dispute Drags On," *Evening Standard*, 24, February 1978, 1.
28. See, for instance, H. Joshua and T. Wallace, *To Ride the Storm. The 1980 Bristol "Riot" and the State*, London: Heineman Educational Press, 1983.
29. S. Cohen and J. Young (eds.), *The Manufacture of News. Deviance, Social Problems and the Mass Media*, London: Constable, 1981.
30. "Teachers Protest over Salary Claims," *Christchurch Star*, June 21, 1978, 24.
31. "Discrimination Say Teachers," *Christchurch Star*, August 23, 1977, 82.
32. Ibid.
33. "Charging of Subject Fees Slated by PPTA," *Christchurch Star*, August 24, 1977, 22.
34. R. Openshaw, *Unresolved Struggle. Consensus and Conflict in State Post-primary Education*, Palmerston North: Dunmore Press, 1995, 113–115.
35. Ibid.
36. Ibid.
37. Ibid.
38. J.G. Jesson, "The PPTA and the State: From Militant Professionals to Bargaining Agent. A Study of Rational Opportunism," PhD diss., University of Auckland, 1995, 187, 197. See also J. Alison, "Mind the Gap! Policy Change in Practice. School Qualifications Reform in New Zealand, 1980–2002," PhD diss., Massey University, Palmerston North, 2007, 97–98.
39. Ibid., 97.

40. W.L. Renwick to Minister of Education, Review of Secondary Curriculum, April 16, 1973. ABEP, W4262, Box 1836 Box 1836, 34/2/13, Part 2. Review of Secondary Curricula.
41. New Zealand Post-Primary Teachers' Association. *Education in Change. Report of the Curriculum Review Group*, Auckland: Longman Paul, 1969, 70.
42. Alison, "Mind the Gap," 102.
43. NZPPTA, *Education in Change*, 2–9.
44. New Zealand Post-Primary Teachers' Association. *The Secondary School Curriculum: 3. The Challenge Is Change*, Auckland: NZPPTA, September 1972, 6.
45. Ibid.
46. Jesson, "The PPTA and the State," 262.
47. K. Sheen to Minister of Education, August 12, 1970, ABEP, W4262. Box 692. 17/1/25, Criticism, Part 2, Archives New Zealand, Wellington.
48. Alison, "Mind the Gap," 99.
49. R. Sweetman, *A Fair and Just Solution? A History of the Integration of Private Schools in New Zealand*, Palmerston North: Dunmore Press in association with the Ministry for Culture and Heritage and the Association of Proprietors of Integrated Schools, 2002, 129.
50. Ibid., 30.
51. Jesson, "The PPTA and the State," 253.
52. See, for instance, J. O'Brien, *A Divided Unity! Politics of NSW Teacher Militancy since 1945*, Sydney: Allen and Unwin in association with New South Wales Teachers' Federation, 1987, 200–205.
53. See, for instance, B. Barry, *Culture and Equality. An Egalitarian Critique of Multiculturalism*, Polity Press, Blackwell: Oxford, 2001.
54. A. Gouldner, *The Future of Intellectuals and the Rise of the New Class*, London: MacMillan, 1979. 1979. See also E. Rata and R. Openshaw (eds.), *Public Policy and Ethnicity. The Politics of Ethnic Boundary-Making*, London: Palgrave Macmillan, 2006.
55. See E. Rata "Goodness and Power: The Sociology of Liberal Guilt," *New Zealand Sociology*, 11, no. 2, November 1996, 225.
56. Gouldner, *The Future of Intellectuals*, 36.
57. Openshaw, *Unresolved Struggle*, 118–122.
58. J. Spring, *Education and the Rise of the Corporate State*, Boston: Beacon Press, 1972. The foreword to this book was written by radical deschooler, Ivan Illich.
59. A. Barcan, *Sociological Theory and Educational Reality. Education and Society in Australia since 1949*, Kensington, NSW University Press, 1993, 61.
60. See, for instance, ibid., especially Chapter 5.
61. See, for instance, C. Vellekoop, "Migration Plans and Vocational Choices of Youth in a Small New Zealand Town," *New Zealand Journal of Educational Studies*, 3, no. 1, 1969, 10–40; and C. Vellekoop, "Streaming and Social Class," *Delta*, 5, August 1969, 12–21.
62. R.J. Bates, "The New Sociology of Education: Directions for Theory and Research," *New Zealand Journal of Educational Studies*, 13, no. 1, May 1978, 3–14.
63. Nash, 1980, 125.

64. See R. Openshaw, G. Lee, and H. Lee, *Challenging the Myths. Rethinking New Zealand's Educational History*, Palmerston North: Dunmore Press, 1993, 276.
65. J. Freeman-Moir, "Enjoyable and Quiet: The Political Economy of Human Misdevelopment," *New Zealand Journal of Educational Studies*, 16, no. 1, May 1981, 17.
66. See J. Codd, R. Harker, and R. Nash (eds.), *Political Issues in New Zealand Education*, first edition, Palmerston North: Dunmore Press, 1985. References to this text, and even photocopied chapters, appear in several Department of Education, Treasury, and SSC files.
67. J. Codd, R. Harker, and R. Nash, "Introduction," in Codd, Harker, and Nash, *Political Issues in New Zealand Education*.
68. D. Harker, "Schooling and Cultural Reproduction," ibid., 69.
69. R. Walker, "Cultural Domination of Taha Maori: The Potential for Radical Transformation," ibid., 77, 81.
70. P. Ramsay, "The Domestication of Teachers. A Case of Social Control," ibid., 103–122.
71. See, for instance, S. Middleton, "American Influences on New Zealand Sociology of Education, 1950–1988," in *The Fulbright Seminars. The Impact of American Ideas on New Zealand's Educational Policy, Practice and Thinking*, edited by J. Philips, J. Lealand, and G. MacDonald, New Zealand–United States Educational Foundation, Wellington: NZCER,1989, 50–55.
72. This had been foreseen in S. Middleton, "Sexual Apartheid or Androgyny? Four Contemporary Perspectives on Women's Education in New Zealand," *New Zealand Journal of Educational Studies*, 17, no. 1, May 1982, 64.
73. For a critique of first-generation American educational revisionism, see D. Ravitch, *The Revisionists Revised. A Critique of Radical Revisionism*, New York: Basic Books, 1977, especially 124–125.
74. G. Dunstall, "The Social Pattern," in *The Oxford History of New Zealand*, edited by G.W. Rice, second edition, Auckland: Oxford University Press, 1992, 481.
75. S. Middleton, "Family Strategies of Cultural Reproduction," in Codd, Harker, and Nash, *Political Issues in New Zealand Education*, 83.
76. B. Yee, "Women Teachers in the Primary Scene: A Study of their Access to Power and Decision-making," MA diss., University of Canterbury, 1985, 44.
77. Openshaw, *Unresolved Struggle*, especially Chapters 2 and 3.
78. See, for instance, ABEP, W4262, Box 1525, 26/1/174/1, Part 1, Education and the Equality of the Sexes—Policy, Archives New Zealand, Wellington.
79. Vocational Guidance and Education for Work: Working Party Recommendations, ABEP, W4262, Box 1525, 26/1/174/1, Part 1, Education and the Equality of the Sexes, Archives New Zealand, Wellington.
80. W.L. Renwick to Minister of Education, January 7, 1977, Women and Education—Policy and Actions, ABEP, W4262, Box 1525, 26/1/174/1, Part 1, Education and the Equality of the Sexes, Archives New Zealand, Wellington.
81. Ibid.
82. R. Openshaw, *Between Two Worlds. A History of Palmerston North College of Education, 1956–1996*, Palmerston North: Dunmore Press, 1996, 85.

83. J. Codd, "Democratic Principles and the Politics of Curriculum Change in New Zealand," in *The Politics of Education in New Zealand*, edited by M. Clark, Wellington: NZCER, 1981, 43.
84. Openshaw, *Unresolved Struggle*, 104.
85. E.A. Gordon, "Access: The Limits and Capacity of the State," DPhil diss., Massey University, Palmerston North, 1989, 116.
86. P. Spoonley, *The Politics of Nostalgia. Racism and the Extreme Right in New Zealand*, Palmerston North: Dunmore Press, 1987, 13.
87. Ibid., 14.
88. A. Ryan, "'For God, Country and Family.' Popular Moralism and the New Zealand Moral Right," MA diss., Palmerston North: Massey University, 1986, 6.
89. R. Openshaw, "Upholding Basic Values: A Case Study of a Conservative Pressure Group," in Codd, Harker, and Nash, *Political Issues in New Zealand Education*, 231.
90. Ibid., 242.
91. Ryan, "For God, Country and Family," 162–163
92. Ryan, "For God, Country and Family," 117.
93. M. Wellington, *New Zealand Education in Crisis*, Auckland: Endeavour Press, 1985, 158; Codd, "Democratic Principles," 53.
94. Wellington, *New Zealand Education*, 77.

4 Only Major and System-wide Reforms Will Suffice

1. S. Webster, *Patrons of Māori Culture. Power, Theory and Ideology in the Māori Renaissance*, Dunedin: University of Otago Press, 1998, 104.
2. R. Piddington, *An Introduction to Social Anthropology*, vol. 1 and 2, Edinburgh: Oliver and Boyd, 1950, 1957, 776. See also ibid., 112.
3. R. Walker, "Maori People since 1960," in *The Oxford History of New Zealand*, second edition, edited by G.W. Rice, Auckland: Oxford University Press, 1992, 502.
4. G. Dunstall, "The Social Pattern," in *The Oxford History of New Zealand*, second edition, edited by G.W. Rice, Auckland: Oxford University Press, 1992, 477.
5. Walker, "Maori People since 1960," 502.
6. T. Robbins, *Cults, Converts and Charisma: The Sociology of New Religious Movements*, London: Sage, 1988.
7. E. Rata, "'Goodness and Power': The Sociology of Liberal Guilt," *New Zealand Sociology*, 11, no. 2, November 1996, 223–276. For a discussion of the role of guilt and confession in capitalist societies, see especially M. Abercrombie, S. Hill, and B.S. Turner, Sovereign Individuals of Capitalism, London, Boston: Allen and Unwin, 1986, 48.

NOTES 199

8. J.L. Hunter, Chairman, National Advisory Committee, memo entitled, "Representation on the National Advisory Committee on Maori Education," undated but c.1969, ABEP, W4262, Box 2898, 46/3/14, National Advisory Committee on Māori Education, Archives New Zealand, Wellington.
9. National Advisory Committee on Maori Education, "Priorities for Developments," July 1970, ABEP, W4262, Box 2848, 53/2/39, Maori, Part 2, Archives New Zealand, Wellington.
10. R. Walker, "Cultural Domination of Taha Maori. The Potential for Radical Transformation," in *Political Issues in New Zealand Education*, edited by J. Codd, R. Harker, and R. Nash, Palmerston North: Dunmore Press, 1985, 75.
11. J.H. Murdoch, *The High Schools of New Zealand: A Critical Survey*, Educational Research Series No. 19, Christchurch: New Zealand Council for Educational Research; R. Openshaw, "Putting Ethnicity into Policy. A New Zealand Case Study," in *Public Policy and Ethnicity. The Politics of Ethnic Boundary-Making*, edited by E. Rata and R. Openshaw, Hampshire and New York: Palgrave Macmillan, 2006, 113–127.
12. M. Simpson, "The Handbook. The Project," c. July 1967, ABEP, Box 2848, 53/2/39, Maori, Part 1, Archives New Zealand, Wellington.
13. J. Bowler, "'The New Zealand Controversy over the Johnson Report,' The Context of the Report of the Committee on Health and Social Education, *Growing, Sharing, Learning* (1977)," PhD diss., Massey University: Albany, New Zealand, 2004.
14. New Zealand Post-Primary Teachers' Association, *Education in Change. Report of the Curriculum Review Group*, Auckland: Longman Paul, 1969, 10; 11.
15. Ibid., 12, 76, 80.
16. E.G. Schwimmer (ed.). *The Maori People in the Nineteen Sixties*, Auckland: Blackwood and Janet Paul, 1968; J. Metge, *The Maoris of New Zealand*, revised edition, London: Routledge and Kegan Paul, 1976, 308.
17. Ibid., 14–15.
18. Ibid., 13.
19. A.F. Smith, "National Advisory Committee on Maori Education, Report of the Office for Māori and Island Education," ABEP, W4262, Box 2898, 46/3/14, National Advisory Committee on Maori Education, Archives New Zealand, Wellington.
20. Report on the Primary National Course Lopdell House, Teachers Colleges, August 30–September 3, 1971, Report entitled "Teachers College Courses Which Prepare Teachers to Provide for the Needs of Maori and Island Children," ABEP, W4262, Box 2898, 46/3/14, National Advisory Committee on Māori Education, Archives New Zealand, Wellington.
21. R. Walker, "The Modern Maori Warden," *Te Maori* 2, July–August 1971, 37.
22. R. Openshaw, "Preparing for Picot: Revisiting the 'Neoliberal' Education Reforms," *New Zealand Journal of Educational Studies*, 38, no. 2, 2003, 135–150.
23. J.G. Jesson, "The PPTA and the State: From Militant Professionals to Bargaining Agent. A Study of Rational Opportunism," PhD diss., University of Auckland, 1995, 231.

24. Bowler, "The New Zealand Controversy."
25. University of Waikato, Centre for Maori Language Studies and Research, *Heads Held High*. Te Kohanga Preschool Project Report for 1974, Working Paper No. 5, December, ABEP, W4262, Box 2848, 53/2/39, Maori, Part 3, Archives New Zealand, Wellington.
26. Openshaw, "Putting Ethnicity into Policy."
27. B. Barry, *Culture and Equality. An Egalitarian Critique of Multiculturalism*, Polity Press, Blackwell: Oxford, 2001; ibid.
28. Metge, *The Maoris of New Zealand*, 329.
29. Ibid., 318.
30. Openshaw, "Putting Ethnicity into Policy."
31. Head Office Submissions to National Advisory Committee on Maori Education, c.1976, ABEP, W4262, Box 2848, 53/2/39, Maori, Part 3, Archives New Zealand, Wellington.
32. Palmerston North Teachers College, Te Kooti Project, 1978, Project proposal dared 4 December, ABEP, W4262, Box 3066, 52/2/53, Life of Te Kooti, Archives New Zealand, Wellington.
33. See, for instance, *The Invention of Tradition*, edited by E. Hobsbawm and T. Ranger, Cambridge: Cambridge University Press, 1983.
34. A.F. Smith, Letter to Bernie Kernot, 15 November, ABEP, W4262, Box 2677, 40/1/4/5, Marae-based Courses for Teachers, Part 1, Archives New Zealand, Wellington.
35. Hauiti Marae, Programme Draft, April 1977, ABEP, W4262, Box 2677, 40/1/4/5, Marae-based courses for teachers, Part 1, Archives New Zealand, Wellington.
36. Marae based courses, Mangamuka, Hokianga, May29–June 2, 1977, Handout. ABEP, W4262, Box 2677, 40/1/4/5, Marae-based Courses for Teachers, Part 1, Archives New Zealand, Wellington.
37. Letter from participant to the organisers of the Te Wai Pounamu College Course, Dunedin, October 31, 1977, ABEP, W4162, Box 2677, 40/1/4/5, Marae-based Courses for Teachers, Part 1, Archives New Zealand, Wellington.
38. Taha Maori across the Curriculum, 1983. No author listed, ABEP, W4262, Box 3732, NS50/2/MA-TM, Maori across the Curriculum, Archives New Zealand, Wellington.
39. Curriculum Development Division, Department of Education, Hui Whakamatatau, Taurua Marae, Rotoiti, November 7–9, 1984, 1, YCDB, A688, 1591a, 33/1, Maori Studies, Part 2, 1985–1987, Archives New Zealand, Auckland.
40. Te Kete Timatatanga. "A Teacher Guide to Taha Maori in Social Studies," Booklet enclosed with a letter from Ash Newth, District Senior Inspector of Primary Schools to nga Rangatira (Principals), October 2, 1985, 6, YCDB, A688, 1591a, 33/1, Māori Studies, Part 2, 1985–1987, Archives New Zealand, Auckland.
41. *Te Kete Timatatanga. A Teacher Guide to Taha Māori in Social Studies* (Ibid.), 7.

42. Taha Maori and Change. Report from National Residential Inservice Course, Lopdell Centre, July 14–18, 1986, YCDB, A688, 1591a, 33/1, Maori Studies, Part 1, 1985–1987, Archives New Zealand, Auckland.
43. Ibid.
44. Ibid.
45. H. Tauroa, *Race Against Time*, Wellington: Human Rights Commission, 1982.
46. Ibid., recommendations 5–7; 56.
47. Ibid., recommendations 9–23; 557–558.
48. Walker, "Cultural Domination of Taha Maori," 77.
49. Ibid.
50. G.H. Smith and L.T. Smith, "Ki Te Whai Ao, Ki Te Ao Marama: Crisis and Change in Māori Education," in *Myths and Realities. Schooling in New Zealand*, edited by A. Jones, G. McCulloch, J. Marshall, H.S. Smith, and L.T. Smith, Palmerston North: Dunmore Press, 1990, 142 (123–156). See also W. Penetito, "Taha Maori and the Core Curriculum," *Delta*, 34, July, 1984, 34–43.
51. Smith and Smith, "Ki Te Whai Ao," 144.
52. Report on Te Kohanga Reo Operations nationally, dated September 1983, entitled: Te Kohanga Reo (The Language Nest). ABRP 6844, W4598, Box 200 31/3/1, Part 6, Te Kohanga Reo-Programme.
53. Anna Jones, for Secretary, Dept of Māori Affairs, March 29, 1983, to District Officers. ABRP 6844, W4598, Box 200 31/3/1, Part 6, Te Kohanga Reo-Programme.
54. Ross for DG to District Senior Inspectors of Primary Schools, August 1, 1983. ABRP 6844, W4598 Box 200 31/3/1, Part 6, Te Kohanga Reo-Programme.
55. Report on Te Kohanga Reo Operations.
56. Ibid., 153.
57. Liz Glasgow, "Shift in Education Needed," in "Today's Woman" Feature, *Auckland Star*, March 28, 1984, B3. See also YCBD, 150a, 33/1 Maori Studies, Part 1, 1978–1984, Archives New Zealand, Auckland.
58. Submission from Citizens Association for Racial Equality. ABEP, W4262. Box 1830, 31/1/64/4, Part 1, Inquiry into Curriculum: Assessment and Qualifications, December 10, 1984.
59. Ibid.
60. Minister of Finance, October 8, 1984. Community Education Initiatives Scheme, 2. AALR, Series 873, W444. Box 247, 52/479/1, Part 2, Archives New Zealand. Wellington. The Minister also attached to this memo a paper from the Minister of Education, with recommendations to Cabinet Committee on Social Equity.
61. Report of the Committee on Gangs, Government Response, AALR, Series 873, W444, Box 247, 52/479/1, Part 2. Archives New Zealand, Wellington.
62. C.E.I.S. Evaluation Report. Otara Resource Network. Interim Report. September 11, 1984. AALR, Series 873, W444. Box 247. 52/479/1, Part 2. Archives New Zealand. Wellington.

63. Minutes of the CEIS Interdepartmental Committee Meeting held on Thursday September 20, 1984, 2. AALR, Series 873, W444. Box 247. 52/479/1, Part 2, Archives New Zealand, Wellington.
64. Ibid., 4.
65. Community Education Initiative Scheme: Meeting with Ministers, Parliament Buildings, October 10, 1984 12.00–1.00pm, 2. AALR, Series 873, W444. Box 247. 52/479/1, Part 2, Archives New Zealand, Wellington.
66. "Care Asks for Commission on Maori Schooling," *Auckland Star*, April 14, 1984, A3.
67. Ibid.
68. W.L. Renwick to Chairman, Waitangi Tribunal, May 28, 1986, ABEP, W4262, 5/1/12. Treaty of Waitangi, Part 1, 1.
69. Ibid., 2.
70. Ibid.
71. Ibid., 3.
72. J. Ennis, "Why Do Our Schools Fail the Majority of Maori Children?" *Tu Tangata*, 36, June/July 1987, 21–22. Ennis had been an inspector of schools for seventeen years.
73. Ibid., 22.
74. J. Ennis, "How Failure Begins in the Classroom," *Tu Tangata*, 36, June/July 1987, 22–23.

5   The Best Kind of Accountability

1. See, for instance, J. Barrington, "Why Picot? A Critique and Commentary," *New Zealand Journal of Educational Administration* 5 (November 1990), 15–17.
2. Ibid.
3. Ibid.
4. See, for instance, ABEP, W4262, Box 1727, 30/2/19, Part 1, English 1945–1982; and ABEP, W4262, Box 3695, NF50/2/ens-la, Part 2, Language across the Curriculum, Archives New Zealand, Wellington.
5. F. Holmes, "Lay and Professional Participation in Educational Administration," Unpublished paper, c. early 1976, BCDQ, 5193R, A739, Lay and Professional Participation in Educational Administration, 1976, Archives New Zealand, Auckland, 16.
6. Barrington, "Why Picot?"
7. *Report of the Advisory Council on Educational Planning* (The Holmes Report), Wellington: Government Printer, 1974.
8. Holmes, "Lay and Professional Participation," 18.
9. Ibid., 20.
10. G. Butterworth and S. Butterworth, *Reforming Education. The New Zealand Experience 1984–1996*, Palmerston North: Dunmore Press, 1998, 17–18.

11. J. Barrington, "Historical Factors for Change in Education," in *Redistribution of Power? Devolution in New Zealand*, edited by P. McKinlay, Wellington: Victoria University Press for Institute of Policy Studies, 1990, 191 (191–213).
12. Ibid., 193.
13. See, for instance, J. Barrington, "The Politics of School Government," in *The Politics of Education in New Zealand*, edited by M. Clark, Wellington: NZCER, 1981, 74.
14. Barrington "Historical Factors for Change in Education," 194.
15. Ibid., 195.
16. See, for instance, P.T. Moody for DG, Head Office Minute 1975/42, Reorganisation of Education Administration, October 15, 1975, BCDQ, 5193N, A739, Reorganisation of Educational Administration 1976. Archives New Zealand, Auckland.
17. N. Scott, *Survey of Some Aspects of New Zealand Secondary Schools*, Wellington. Department of Education, 1977.
18. Ibid., 4.
19. Ibid., 16.
20. OECD Review of New Zealand Educational Policies 1982, Draft Statement for the Examining Panel. Department of Education. Wellington, 1982 February. See also, New Zealand Department of Education, *Education Policies in New Zealand: Report Prepared for the OECD Examiners by the Department of Education in March 1982*, Wellington, 1983.
21. Ibid., 27.
22. Ibid., 22–29.
23. Ibid., 43–45.
24. Ibid., Introduction.
25. OECD, *Review of National Policies for Education: New Zealand*. Paris, 1983, 27–28. See also Butterworth and Butterworth, *Reforming Education*, 19–20.
26. Barrington, "The Politics of School Government," 82–84.
27. Barrington, "Historical Factors for Change in Education," 196.
28. Ibid., 198.
29. *NZPD*, vol. 469, 1985 (March 26), 691.
30. P.A. Atkinson, to DG, Assistant DGs, Assistant secretaries, National Superintendents of Education, and all Directors, entitled "Revision of the Education Act—Progress Report," July 17, 1986, BCDQ, 4031D, A739, 1/15. Revision of Education Act, 1.
31. Ibid., 2.
32. Ibid.
33. Butterworth and Butterworth, *Reforming Education*, 66.
34. Ibid., 65–66.
35. Ibid., 66–67.
36. "'Aberration' Needs Study," *Evening Post*, April 11, 1987.
37. New Zealand Department of Education, *The Curriculum Review*. Report of the Committee to Review the Curriculum for Schools, Wellington, 1987.
38. Ibid., 8–9.
39. Ibid., 34.

40. Ibid., 36.
41. Ibid., 37.
42. Ibid., 37.
43. "Education Reshuffle," *Dominion,* June 24, 1987, editorial, 14.
44. Ibid.
45. Rt Hon David Lange to PPTA Conference, Sheraton Hotel, Auckland, August 25, 1987, Speech notes, AALR 73, W5427, Box 1064, 62/9, Part 27 Education Miscellaneous. See also D. Dawson, "Education Response of Concern to Lange," *Dominion,* August 26, 1987, 7.
46. Lange, PPTA Speech notes, 4.
47. Ibid., 8–10.
48. Ibid., 16.
49. Ibid., 21.
50. Rt Hon David Lange to the New Zealand Secondary School Boards Association Biennial Conference, Central Institute of Technology, Trentham, Speech notes. August 28, 1987, AALR 73, W5427, Box 1064, 62/9, Part 27, Education Miscellaneous.
51. Ibid., 2.
52. Ibid., 8.
53. Ibid., 13–14
54. Ibid., 16.
55. Ibid.
56. Ibid., 18.
57. Barrington, "Historical Factors for Change in Education," 198.
58. *Vocational Training Council. Education and the Economy.* A Report prepared for the Organisation for Economic Co-Operation and Development (OECD) by the New Zealand Vocational Training Council. June 1986.
59. Butterworth and Butterworth, *Reforming Education,* 66–67.
60. M. McKinnon, *Treasury. The New Zealand Treasury 1840–2000,* Auckland: Auckland University Press, 2003, 274.
61. M.K. Burns for DG, to Minister of Education, September 13, 1979, ABEP. W4262, Box 485, 1/2/3, Financial and Resource Management, Part 1, Archives New Zealand, Wellington.
62. The Treasury, Task Force Report: New Machinery for Public Sector Planning, The Treasury: Wellington, August 10, 1977 (unpublished), ABEP, W4242, Box 485, 1/2/3, Financial and Resource Management, Part 2.
63. N.F. Sutton to DG. Re-Task Force Report: New Machinery for Public Sector Planning, August 16, 1977, ABEP, W4242, Box 485, 1/2/3, Financial and Resource Management, Part 2.
64. Ibid.
65. The Report of the Controller and Auditor General on Financial Management in Administrative Government Departments (unpublished). The Treasury: Wellington, May 1978 ABEP, W4242, Box 485, 1/2/3, Financial and Resource Management, Part 2.
66. Financial Management and Review Team, Review of Financial Management and Control. The Treasury: Wellington, April 1978, ABEP, W4262, Box 485, 1/2/3, Financial and Resource Management, Part 1.

67. Political Reporter, "Where Your Money Should NOT Have Gone," *Dominion*, June 28, 1978, 1.
68. W.L. Renwick to Minister of Education, July 25, 1978, ABEP, W4262, Box 485, 1/2/3, Financial and Resource Management, Part 1.
69. Ibid.
70. C.H.S. Miller for DG, to Associate Minister of Finance, September 20, 1978, ABEP, W4262, Box 485, 1/2/3, Financial and Resource Management, Part 3.
71. C.H.S. Miller for DG to Secretary of the Treasury, November 9, 1978, ABEP, W4262, Box 482, 1/2/2, Miscellaneous Correspondence.
72. W.L. Renwick, Comments on the Treasury Report re—the Development of the Psychological Service March 28, 1979, ABEP, W4262, Box 485, 1/2/3, Financial and Resource Management, Part 3.
73. M.K. Burns for DG, to Minister of Education, September 13, 1979, ABEP, W4262, Box 485, 1/2/3, Financial and Resource Management, Part 1.
74. N.F. Sutton for DG to Secretary for the Treasury, February 21, 1980, ABEP, W4262, Box 485, 1/2/3, Financial and Resource Management, Part 1.
75. Secretary of Treasury to Education Department Accountant, September 17, 1979, ABEP, W4262, Box 485, 1/2/3. Financial and Resource Management, Part 2.
76. D.F. Quigley to Chairman, Cabinet Committee on Expenditure, March 14, 1980, 2, ABEP, W4262, Box 485. 1/2/3. Financial and Resource Management, Part 2.
77. Ibid., 5.
78. W.L. Renwick to Minister of Education, May 19, 1980, ABEP, W4262, Box 485, 1/2/3, Financial and Resource Management. Part 2.
79. M.K. Burns for DG, to Minister of Education, August 11, 1980, ABEP, W4262, Box 485, 1/2/3, Financial and Resource Management. Part 2.
80. See, for instance, B. Lagan, "Spending Cuts Target Higher," *Dominion*, March 5, 1982, 1; "Expenditure Cuts Could Reach 8 1/2 pc," *Evening Post*, March 5, 1982, Section 1, 4.
81. Notes of Minister's meeting with Education Board Chairmen held on February 4, 1982, ABEP, W4262, Box 485, 1/2/8, Budget Preparation, Part 9.
82. S. Birks and S. Chatterjee, *The New Zealand Economy. Issues and Policies*, second edition, Palmerston North: Dunmore Press, 1992, 240–241.
83. See, for instance, "Cut Condemned by Inspectors," *Evening Post*, March 9, 1982, 27; "Teacher Says Cuts Add to Rejects," *Dominion*, March 9, 1982, 3.
84. See, for instance, *Education and the Economy*, A Report prepared for the Organisation for Economic Co-Operation and Development (OECD) by the New Zealand Vocational Training Council, June 1986, ABEP, W4262, Box 3634, NS41/9/3, EWE.
85. See, for instance, Commission of Inquiry into an Alleged Breach of Confidentiality of the Police File on the Honourable Colin James Moyle, MP, Wellington: Prime Minister's Department Press Office, 1978.
86. Report of the Royal Commission to Inquire into the Circumstances of the Conviction of Arthur Allan Thomas for the Murders of David Harvey Crewe

and Jeanette Lenore Crewe, 1980. The police came in for considerable public and official scrutiny over the next few years. See, for instance, *Inquiry into Reported Allegations of Police Misuse of the Wanganui Computer. Report of the Wanganui Computer Centre Privacy Commissioner*, Wellington: Government Printer, May 1986.
87. Report of the Royal Commission to Inquire into the Crash of a DC10 Aircraft operated by Air New Zealand Limited, 1981. Presented to the House of Reps by Command of His Excellency the Governor-General (Sir Keith Holyoake). Wellington: Government Printer.
88. Foreign Affairs and Defence Select Committee. Report of the Inquiry into Disarmament and Arms Control, 1985, Wellington: Government Printer, 15–16; 20–21.
89. *Metro*, June 1987, 4–65
90. The Report of the Committee of Inquiry into Allegations concerning the Treatment of Cervical Cancer at National Women's Hospital and into Other Related Matters, Auckland: Government Printer, 1988, July, 210–218.
91. *Unshackling the Hospitals*, Report of the Hospital & Related Services Taskforce, Wellington, 1988, 11.
92. Ibid., 12.
93. Ibid., 13.
94. Ibid., 26.
95. Education and Science Select Committee. *Report of the Inquiry into the Quality of Teaching*, Second Session (The Scott Report), Wellington: Government Printer, 1986, 1.6.5, 7.
96. Ibid.
97. Ibid., 47.
98. Ibid., 24.
99. Ibid., 28.
100. Ibid., 37.
101. Ibid., 28–34.
102. Ibid., 36.
103. Ibid., 36–37.
104. R. Fargher and M. Probine, *The Report of a Ministerial Working Party. The Management, Funding and Organisation of Continuing Education and Training*, Wellington, 1987 (March), i.
105. Ibid.
106. Ibid., ii.
107. Ibid., ii.

6 A Blank Page Approach

1. R.O. Douglas, Minister of Finance, to Russell Marshall, Minister of Education, June 9, 1987, Memo entitled "Review of the State Education

Service," ABEP, W4262, Box 923, 17/5/5, Part 1, Review of Educational Administration.
2. Transcript of interview 33 with Maurice Gianotti, September 10, 1996, Wellington, G. Butterworth and S. Butterworth, Ministry of Education Oral History Project. Education Reform 1987–1995, 5.
3. Transcript of interview 29 with Brian Picot, November 1, 1995, Auckland, G. Butterworth and S. Butterworth, Extracts from Ministry of Education Oral History Project. Education Reform 1987–1995, 1.
4. Picot Video Script, AAWW, 7112, W4262, Box 5, National Library, Picot, Archives New Zealand, Wellington.
5. Report of the Taskforce to Review Education, *Administering for Excellence. Effective Administration in Education* (The Picot Report), Wellington: Government Printer, 1988, ix
6. Ibid., ix.
7. Ibid., ix.
8. M. Gianotti, Notes of a Meeting of the First Meeting of the Taskforce Held in Wellington July 31, 1987, 3. AAZY W3901, Box 181, Picot—General, Archives New Zealand, Wellington.
9. Ibid., 3.
10. Transcript of interview 33 with Maurice Gianotti, 8.
11. Taskforce to Review Educational Administration, Meeting 8.00 am to 4.30 pm, Friday July 31, 1987, AAZY W3901, Box 181, Picot—General, Archives New Zealand, Wellington.
12. Ibid., 7.
13. Ibid., 2.
14. Transcript of interview 33 with Maurice Gianotti.
15. Transcript of interview 29 with Brian Picot, 8.
16. Transcript of interview 33 with Maurice Gianotti, 5.
17. Meeting 8.00am to 4.30pm, Friday July 31, 1987.
18. Ibid.
19. Ibid., 7.
20. See item 79, Folder 15, untitled thirty page document providing abstracts of readings for the Picot Taskforce, AAZY, W.3901, Box 181, Picot-General.
21. Meeting 8.00am to 4.30pm, Friday July 31, 1987, 3.
22. Ibid., 3–4.
23. Ibid., 4.
24. Ibid.
25. Ibid., 5.
26. Ibid., 4.
27. Ibid., 6–7.
28. Ibid., 6.
29. L. Webb, *The Control of Education in New Zealand*, Auckland: New Zealand Council for Educational Research, 1937.
30. Task Force to Review Education Administration, Record of Policy Points Made and Tasks to Be Undertaken from Meeting Number Two of August 17,

1987, AAZY, W3901, Box 181, Picot—General, Archives New Zealand, Wellington.
31. Ibid.
32. Ibid.
33. Ibid.
34. Task Force Status Report 31.8.87, AAZY W3901, Box 181, Picot—General, Archives New Zealand, Wellington.
35. The Treasury, *Government Management. Brief to the Incoming Government 1987*, vol. 1 (untitled), Wellington: Government Printer, August.
36. Ibid., 358–359; 368.
37. Ibid., 132–134.
38. The Treasury. *Government Management. Brief to the Incoming Government 1987*, vol. 11, Education Issues, Wellington: Government Printer, August, preface.
39. Ibid., 10.
40. Ibid., 140–142.
41. Ibid., 152.
42. Transcript of interview 5 with Simon Smelt, Chief Analyst with Treasury, March 23, 1994, Wellington, G. Butterworth and S. Butterworth, Ministry of Education Oral History Project. Education Reform 1987–1995.
43. *Government Management*, vol. 11, 243.
44. Ibid., 244.
45. J.H. Murdoch, *The High Schools of New Zealand: A Critical Survey*, Christchurch: NZCER, 1944, 429–430.
46. See, for instance, *Government Management*, vol. 11, 272.
47. Draft Submission to Taskforce on Education Administration (Department of Education), undated but c. October 1987, AAWW, Box 9, Education including Picot Taskforce, Archives New Zealand, Wellington, 7.
48. Ibid., 13, point 42.
49. Ibid., 23, point 72.
50. Transcript of interview 29 with Brian Picot, 11.
51. Transcript of interview 33 with Maurice Gianotti, 6.
52. "PM Wants to Axe Dept of Education," *Evening Post*, November 7, 1987, 1.
53. "PM wants to axe Dept of Education" (Ibid.), Front page lead, 1.
54. Administration of Education, Underlying Efficiency Considerations, Outline Note by Treasury, no page nos., AAWW, Box 9, Education including Picot Taskforce, Archives New Zealand, Wellington.
55. Taskforce to Review Education Administration, Possible Organisational Models, Note by Treasury (Undated, no author supplied), AAWW, Box 9, Education Including Picot Taskforce, Archives New Zealand, Wellington.
56. S.J. Smelt, Accountability in Education. Note by Secretariat to Taskforce to Review Education Administration, November 9, 1987, AAWW, Box 9, Education Including Picot Taskforce, Archives New Zealand, Wellington.
57. W.D. Scott Deliotte Ltd., *Report to the Taskforce to Review Education Administration. Initial Assessment of the Functions of the Department of Education*. November 1987, Introduction, 1.

58. Ibid., Executive Summary, 2.
59. Ibid., Executive Summary, 1.
60. Ibid., 55.
61. See ibid., Folder 15.
62. Government Research Unit, Education Vouchers, February 4, 1987, Folder 15, AAZY, W3901, Box 181, Picot-General, Archives New Zealand, Wellington.
63. New Zealand Council for Educational Research, *How Fair Is New Zealand Education?* Part 1, NZCER for Royal Commission on Social Policy, Wellington, 1987, 1. The accompanying footnote explained that this breakdown was based on 50% girls and women, 30–35% low SES, 15% Māori, 5% disabled and 13% rural, which was claimed to be "a very conservative view of the degree of overlap between these groups."
64. Ibid., 5.
65. R. Benton, *How Fair Is New Zealand Education? Part 11. Fairness in Māori Education. A Review of Research and Information.* The Royal Commission on Social Policy: NZCER. Wellington, November 1987.
66. Ibid., 69.
67. Ibid., 71.
68. Ibid., 71.
69. K. Polanyi, *The Great Transformation*, Farrar & Rinehart: New York and Toronto, 1944, 3–4.
70. R, Sandall, *The Culture Cult. Designer Tribalism and Other Essays*, Boulder, CO: Westview Press, 2001, ix, 88.
71. L. Moss, "Picot as Myth," in *Critical Perspectives: New Zealand Education Policy Today*, edited by S. Middleton, J. Codd, and A. Jones, Wellington: Allen and Unwin in association with Port Nicholson Press, 1990, 145.
72. Ibid., 145.
73. W.D. Scott Deliotte Ltd., Folder 15, item 29, 8.
74. Transcript of interview 33 with Maurice Gianotti, 8.
75. A. Dixon to PM, February 15, 1988, Picot Taskforce—Progress Report, AAZY, W3901, Box 181, Picot-General, Archives New Zealand, Wellington.
76. Transcript of interview 33 with Maurice Gianotti, 8.
77. M. Bazley, The Watts Report on New Zealand Universities and Post-Compulsory Education and Training, Social Services Division, SSC, Wellington, February 22, 1988 (6 pp.), AAZY, W3901, Box 181, Picot-General, Archives New Zealand, Wellington.
78. Ibid., 3.
79. R.J.S. Macpherson, Report on Structural Changes in Victoria's Education, SSC, March 2, 1988, AAZY, W3901, Box 181, Picot-General, Archives New Zealand, Wellington. Copies of this report were also sent to the Prime Minister's Office, SSC, Treasury, the Department of Labour, and the Assistant Minister of Education.
80. Ibid., Appendix 6, 1.
81. Ibid., Appendix 6, 7.
82. Ibid., 1–2.

83. Ibid., 3.
84. Ibid., 4–5.
85. Picot Report, xi.
86. Ibid., 6.
87. Ibid., 22.
88. Ibid., 23–24.
89. Ibid., xi.
90. Transcript of interview 20 with Whetu Wereta, June 20, 1994, Wellington, G. Butterworth and S. Butterworth, Ministry of Education Oral History Project. Education Reform 1987–1995.
91. Picot Report, xi.
92. Ibid., xii.
93. Transcript of interview 33 with Maurice Gianotti, 10.
94. Transcript of interview 29 with Brian Picot, 5.
95. Picot Report, xii.
96. Ibid., xiii.
97. Ibid., xiii.
98. Ibid., 17.
99. Ibid., xiv.
100. Ibid., 17.
101. Ibid., 18.
102. Ibid., 17.

7 A Long Way to Go before We Win the Battle

1. See, P.J. Gibbons, "The Climate of Opinion," in *The Oxford History of New Zealand*, edited by G.W. Rice, Auckland: Oxford University Press, 1992, 308–336. See also New Zealand Department of Education, *OECD Review of New Zealand Educational Policies 1982. Draft Statement for the Examining Panel*, New Zealand Department of Education, Wellington, February 1982, 27.
2. D. Lange, Education Media Briefing, April 13, 1988, AALR, 873, W5427/164, 62/9, Part 28, 3, Archives New Zealand, Wellington.
3. Lange (Ibid.), 4.
4. H. McQueen, The Year of Educational Change: A Strategy to implement the Picot Report and enhance the Government's Image in Education, c. February 1988, AAWW, Series 7112, W4640, Box 9, Education including Picot Taskforce, Archives New Zealand, Wellington.
5. Ibid.
6. Ibid.
7. Ibid.
8. Ibid.

9. H. McQueen, Memo to John Henderson, March 2, 1988, AAWW, Series 7112, W4640, Box 9, Education Including Picot Taskforce, Archives New Zealand, Wellington.
10. McQueen, February 1988.
11. Cited in J. Bowler, "The New Zealand Controversy over the Johnson Report: The Context of the Report of the Committee on Health and Social Education, *Growing, Sharing, Learning* (1977)," PhD diss., Massey University, 2005, 364.
12. A. Ross, Acting DG, to S. Logan, Managing Director, Logos Public Relations Ltd, April 29, 1988, AAWW, Series 7112, W4640, Box 4: Logos/Picot, Curriculum Review, Archives New Zealand, Wellington.
13. Ibid.
14. Picot Implementation Phase Budgets and General Strategy and Submission Analysis, AAWW, Series 7112, W4640, Box 4: Logos/Picot, Curriculum Review, Archives New Zealand, Wellington.
15. Minister of Education Releases Picot Report, Press statement dated May 10, 1988, AAWW. Series 7112, W4640, Box 4: Logos/Picot, Curriculum Review, Archives New Zealand, Wellington.
16. "Now for a Lesson in Education," *Evening Post*, May 11, 1988, 6, editorial.
17. "Education Will Never Be the Same Again," *The Star*, May 11, 1988, 8, editorial.
18. "The Picot Report," *The Press*, May 12, 1988, 12, editorial,
19. "Picot Report Backed," *Northern Advocate*, May 11, 1988, 2, editorial.
20. Confidential letter from S. Logan, Managing Director, to H. McQueen, PMs Department, June 8, 1988, AAWW, Series 7112, W4640. Box 4: Logos/Picot, Curriculum Review, Archives New Zealand, Wellington.
21. Ibid.
22. Outline Approach to Implementation Phase June 7, 1988, AAWW, Series 7112, W4640, Box 4: Logos/Picot, Curriculum Review, Archives New Zealand, Wellington.
23. Ibid., 1–3.
24. Ibid., 10.
25. G. Hubbard, "'New Right' Threat to Teacher Unions," *Waikato Times*, May 18, 1988, 3.
26. R. Fry, "Feature: School and Community. The Partnership," *PPTA Journal*, Term 2, 1988, 1, editorial.
27. R. Nash and W. Korndorffer, "After Picot Community Education will be Something Else," *PPTA Journal*, 2, 1988, 3.
28. Ibid., 3–4.
29. "National Offers Support," *Evening Post*, May 11, 1988, 2; Newzel Log: Radio New Zealand National Programme, May 14, 1988, Transcript.
30. J. Long, education reporter, "Tighter Control for Schools over Funds?" *The Press*, June 9, 1988, 8.
31. Ibid.
32. D. Lange, to Chair of Cabinet Social Equity Committee. Picot Taskforce: Officials Committee, April 11, 1988, AAWW, Series 7112, W4640, Box 24, Social Equity, Archives New Zealand, Wellington.

33. Responses to the Picot Report, July 13, 1988, AAWW, Series 7112, W4640, Box 24, Social Equity, Archives New Zealand, Wellington.
34. Responses to the Picot Report (Ibid.), 7.
35. Logos, Letter Analysis, June 16, 1988, AAWW, Series 7112, W4640, Box 24, Social Equity, Archives New Zealand, Wellington.
36. See "Summary of Sector Responses," in Responses to the Picot Report.
37. Ibid.
38. Ibid.
39. Submission from Nga Iwi Māori—from the Māori People throughout New Zealand, tabled by Wiremu Kaa, July 1, 1988, AAWW, Series 7112, W4640, Box 24, Social Equity, Archives New Zealand, Wellington.
40. Ibid.
41. Ibid.
42. Ibid.
43. G. Knight J.A. Ross, Picot Report and the Integration Act, July 3, 1988, AAWW, Series 7112, W4640, Box 24, Social Equity, Archives New Zealand, Wellington.
44. Report of the Officials Committee on the Reform of Educational Administration: Revised Paper, July 25, 1988, AAWW, Series 7112, W4640, Box 24, Social Equity, Archives New Zealand, Wellington.
45. Ibid., 40.
46. Ibid., 51.
47. Ibid., 18.
48. The Picot Report Survey, Detailed Tabular Report, AAWW, Series 7112, W4640, Box 4: Logos/Picot, Curriculum Review. See also covering fax by Logos dated July 25, 1988.
49. Ibid.
50. National Survey—Key Findings and Recommendations, dated July 26, 1988, Insight New Zealand, AAWW, Series 7112, W4640, Box 4: Logos/Picot, Curriculum Review, Archives New Zealand, Wellington.
51. Ibid.
52. Ibid., 1.
53. Ibid., 1–2.
54. Ibid., 1–2. This latter danger was rendered more acute when, to the doubtless chagrin of the Prime Minister's Office and the embarrassment of Logos, the failure of a Logos employee to return confidential survey questionnaires on school committees resulted in the contract with Logos being abruptly terminated in August 1988.
55. J. Brame, "Towards Supermarket Education," *Broadsheet*, no. 160, July/August 1988, 9.
56. Ibid.
57. A. Levett, "Education Reviews Fall Short," *National Business Review*, July 13, 1988, 6.
58. "Trim PM Tackles 'New Right,'" *New Zealand Herald*, July 25, 1988, 9.
59. "Professor Reveals Pressure Put on the Picot Report," *Herald*, July 25, 1988, 5; "Picot Report Member Changes His Mind," *Evening Post*, July 25, 1988, 1.

60. Briefing notes for Press Conference, August 7, 1988, AAWW, Series 7112, W4240, Box 19, Teacher HR, Picot Implementation, Archives New Zealand, Wellington.
61. *Tomorrows' Schools. The Reform of Education Administration in New Zealand*, Wellington: Government Printer, August 1988.
62. Ibid., 25.
63. Transcript of interview 29 with Brian Picot, November 1, 1995, Auckland. G. Butterworth and S. Butterworth, Ministry of Education oral History Project. Education Reform 1987–1995.
64. *Tomorrows' Schools*, 7–8.
65. Ibid., especially 1.3.1, 26.
66. Transcript of interview 5 with Simon Smelt, March 23, 1994, Wellington. G. Butterworth and S. Butterworth, Ministry of Education oral History Project. Education Reform 1987–1995.
67. Transcript of interview 20 with Whetu Wereta, June 20, 1994, Wellington. G. Butterworth and S. Butterworth, Ministry of Education oral History Project. Education Reform 1987–1995.
68. *Tomorrows' Schools*, 26.
69. Ibid., 26.
70. See chapter 8.
71. See, for instance, P. Ellison, *The Manipulation of Māori Voice: A Kaupapa Māori Analysis of the Picot Policy Process*, Department of Education. Institute for Research and Development in Māori Education, Victoria University, Wellington, 1997.
72. R. Fry, Editorial. *PPTA Journal*, Term 3, 1988, 1.
73. J. Codd, "Picot: A Risky Reform?" *PPTA Journal*, Term 3, 1988, 2–5.
74. C. Lankshear, "The Picot Report as a Cultural Nightmare," *PPTA Journal*, Term 3, 1988, 8.
75. G. Hingangaroa-Smith, "The Picot Report: A Cocktail for a Cultural and Social Catastrophe," *PPTA Journal*, Term 3, 1988, 17–18.
76. P. Capper, "Wrong Questions—Wrong Answers," *PPTA Journal*, Term 3, 1988, 11.
77. Ibid., 12.
78. A. Dalziell, "Tomorrow's Schools from a Rural Perspective," *PPTA Journal*, Term 3, 1988, 21–23; H. Watson, "The Fate of Women and Girls in Tomorrow's Schools," *PPTA Journal*, Term 3, 1988, 29–30.
79. Dalziell, "Tomorrow's Schools," 21.
80. Watson, "Fate of Women and Girls," 30.
81. R. Nash, "The Treasury on Education: Taking a Long Spoon," *New Zealand Journal of Educational Studies*, 23, no. 1, 1988, 40, 43.

8 Only Half a Policy

1. M. Lange to David Lange, March 21, 1989, AAWW, Series 712, Box 9, Education including Picot Taskforce, Archives New Zealand, Wellington.

2. Memo entitled, "Ministry of Education Setting Up (as at 21 November 1988)," BCDQ, A739 4227a, 1/17/1, New Ministry of Education, Archives New Zealand, Auckland.
3. P. Goff, Associate Minister of Education, "Memorandum to Cabinet Social Equity Committee, Ministry of Education," undated but c. late 1989, AALR, 873, W5427, Box 1065, 62/9, Part 29, Education—Miscellaneous, Archives New Zealand, Wellington.
4. "New Release from the State Services Commission," July 4, 1989, BCDQ, A739, 4227a, 1/17/1, New Ministry of Education, Archives New Zealand. Auckland.
5. J. Codd and L. Gordon, "School Charters: The Contractualist State and Education Policy," *New Zealand Journal of Education Studies*, 26, no. 1 (1991), 21–34.
6. Document entitled T89/140 of September 13, 1989, Appended to P. Goff, Memorandum to Cabinet Social Equity Committee, Ministry of Education, AALR 873, W5427, Box 1065, 62/9, Part 29, Education—Miscellaneous, 8.
7. P. Goff, "Memorandum to Cabinet Social Equity Committee," 8.
8. *Today's Schools. A Review of the Education Reform Implementation Process* (The Lough Report), Prepared for the Minister of Education, April 1990.
9. Ibid., 6.
10. Ibid., 9.
11. Ibid., 8–9.
12. P. Carpinter, "Principles Underlying the State Sector; the SSD View," September 20, 1990 (12 pp.), AAFH, Series 6790, W5510, Box 438, SS-1-3-1-X, Part 1, General Policy, Archives New Zealand, Wellington.
13. Ibid., 2.
14. Ibid., 5.
15. Ibid., 8.
16. D.K. Hunn, "Overview of Key Issues," October 29, 1990, 1, AAFH, Series 6790, W5510, Box 438, SS-1-3-1-X, Part 1, General Policy, Archives New Zealand, Wellington.
17. Ibid., 2.
18. Ibid., 4.
19. Ibid., 7.
20. G. Butterworth and S. Butterworth, *Reforming Education. The New Zealand Experience 1984–1996*, Palmerston North: Dunmore Press, 1998, 179–180.
21. S. Sexton, *New Zealand Schools. An Evaluation of Recent Reforms and Future Directions*, Wellington: New Zealand Business Roundtable, December 1990.
22. Ibid., 16.
23. Ibid., 18.
24. Ibid., 31.
25. Ibid., 33.
26. "The Major Issues for Education, Science and Technology Official's Committee," February 4, 1991, AAFH, Series 6790, W5510, Box 458, SS-1-3-4, Part 1,

Structures and Systems—Social Policy—Education—Schools—Policy, 1–2, Archives New Zealand, Wellington.
27. Ibid., 2.
28. Ibid., 3.
29. Ibid., 4.
30. Ibid., 4.
31. W.F. Birch, Memorandum for Cabinet Expenditure Control Committee entitled, "Today's Schools—Report on the Ministry of Education's Outputs," March 1, 1991 (3 pp.). AAFH, Series 6790, W5510, Box 458, SS-1-3-4, Part 1, Structures and Systems—Social Policy- Education—Schools—Policy, 1–2, Archives New Zealand, Wellington.
32. D. Greig, for Secretary to the Treasury to Minister of Finance, Memo entitled, "Ministry of Education: Size of Outputs," March 4, 1991 (4 pp.), AAFH, Series 6790, W5510, Box 458, SS-1-3-4, Part 1, Structures and Systems—Social Policy—Education—Schools—Policy, 4, Archives New Zealand, Wellington.
33. Ibid., 1.
34. Ibid., 1–2.
35. C. Gibson, Group Manager Operations, for Review Team (Ministry of Education), to Treasury, March 6, 1991, AAFH, Series 6790, W5510, Box 458, SS-1-3-4, Part 1, Structures and Systems—Social Policy—Education—Schools—Policy, 4, Archives New Zealand, Wellington.
36. Cabinet Expenditure Control Committee, Minutes of a Meeting held on March 12, 1991. AAFH, 6790, W5510, Box 458, SS-1-3-4, Part 1, Structures and Systems—Social Policy—Education—Schools—Policy, 2, Archives New Zealand, Wellington.
37. Ibid., 3.
38. Cabinet Expenditure Control Committee, "Education Expenditure," March 15, 1991. AAFH, 6790, W5510, Box 458, SS-1-3-4, Part 1, Structures and Systems—Social Policy—Education—Schools—Policy, 4, Archives New Zealand, Wellington.
39. See State Services Commission, Learning Media Review, March 20, 1991, AAFH, Series 6790, W5510, Box 458, SS-1-3-4, Part 1, Structures and Systems—Social Policy—Education—Schools—Policy, Archives New Zealand, Wellington. See also, Review Committee Report, Learning Media, April 10, 1991, AAFH, Series 6790, W5510, Box 458, SS-1-3-4, Part 1, Structures and Systems—Social Policy—Education—Schools—policy, Archives New Zealand, Wellington.
40. Inter Office Memorandum from D. Tripp to S. Hitchiner and D. Greig entitled "Ministry Outputs," March 13, 1991, AAFH, Series 6790, W5510, Box 458, SS-1-3-4, Part 1, Structures and Systems—Social Policy—Education—Schools—Policy, Archives New Zealand, Wellington.
41. J. Gill to D. Greig, May 24, 1991. AAFH, Series 6790, W5510, Box 458, SS-1-3-4, Part 1, Structures and Systems—Social Policy—Education—Schools—Policy, Archives New Zealand, Wellington.

42. D. Greig to D. Tripp, May 2, 1991, AAFH, Series 6790, W5510, Box 458, SS-1-3-4, Part 1, Structures and Systems—Social Policy—Education—Schools—Policy, Archives New Zealand, Wellington.
43. H. Lauder, *The Lauder Report. Tomorrow's Education, Tomorrow's Economy*, a report commissioned by the Education Sector Standing Committee of the New Zealand Council of Trade Unions, Wellington, June 1991.
44. Ibid., 3.
45. Ibid., 11.
46. G. Crocombe, M.J. Enwright, and M.E. Porter, *Upgrading New Zealand's Competitive Advantage*, Auckland, Melbourne, Oxford and New York: Oxford University Press, 1991, Foreword, 8.
47. Ibid., 153.
48. Ibid., 168.
49. L.M. Smith, A Strategy for Investment in Education and Training, draft 3, May 1991, addressed to Chair, Cabinet Strategy Committee, Executive Summary,1, AAFH, Series 6790, W 5510, Box 438, SS-1-3-1-X, Part 1, General Policy, Archives New Zealand, Wellington.
50. Ibid., 5.
51. Ibid., 6.
52. New Zealand Ministry of Education, *The New Zealand Curriculum Framework*, Wellington: Learning Media, 1993, 1.
53. Cited in G. Lee and D. Hill, "Curriculum Reform in New Zealand: Outlining the New or Restating the Familiar?" *Delta*, 48, no. 1, 1996, 29.
54. Ibid., 30.
55. National Curriculum, Key Issues Requiring Immediate Development, c. early 1992, no author supplied. Private Papers (held by author).
56. New Zealand Ministry of Education *The New Zealand Curriculum Framework*, 28.
57. R. Openshaw, "'Citizen Who?' The Debater over Economic and Political Correctness in the Social Studies Curriculum," in *New Horizons for New Zealand Social Studies*, edited by P. Benson and R. Openshaw, Massey University: ERDC Press, 1998, 19–42.
58. New Zealand Qualifications Authority, *Learning to Learn. An Introduction to the New Zealand Qualifications Framework*, <Place>: NZQA, 1992, 3.
59. New Zealand Ministry of Education, White Paper entitled, *The National Qualifications Framework of the Future*, Wellington: Ministry of Education, 1999.
60. New Zealand Ministry of Education, *Education for the Twenty-first Century*, Wellington: Learning Media, 1994, Foreword by L.M. Smith.
61. Comments c. March 1991, AAFH, Series 6790, W5510, Box 452, ss-1-3-4-1, Part 1, Structures and Systems—Social Policy—Education, Archives New Zealand, Wellington.
62. C. Pilgrim, STA Special Report Rural Schools. The Fabric of the Rural Community, March 1991, AAFH, Series 6790, W5510, Box 452, ss-1-3-4-1, Part 1, Structures and Systems—Social Policy—Education, Archives New Zealand, Wellington.

63. State Services Commission Surveys, Boards of Trustees, c. March 1991, AAFH, Series 6790, W5510, Box 452, ss-1-3-4-1, pt 1. Structures and Systems—Social Policy—Education, Archives New Zealand, Wellington.
64. Main points from School Visits, c. March 1991, AAFH, Series 6790, W5510, Box 452, ss-1-3-4-1, Part 1, Structures and Systems—Social Policy—Education, Archives New Zealand, Wellington.
65. Extracts from conversation between Graye Shattky (G) and Lockwood Smith (L)—Kou Lounge, Auckland, March 2, 1991, 6 pm to 7.25 pm, AAFH, Series 6790, W5510, Box 452, ss-1-3-4-1, Part 1, Structures and Systems—Social Policy—Education, Archives New Zealand, Wellington.
66. Ibid.
67. Ibid.
68. *Tomorrows' Schools. The Reform of Education Administration in New Zealand*, Wellington: Government Printer, August 1988, 14.
69. *NZPD*, vol. 542, 1994, 3405.
70. K. Brown, Education Reporter, "Post Exclusive; Sell Off Schools, Treasury Suggests," *Evening Post*, May 10, 1991, 1.
71. See, for instance, "Treasury Denies Sale," *Evening Post*, May 11, 1991, 1.
72. J. Anderton, "'Output-Based' Advice from the Treasury Factory to Sell Schools," Press Release, May 13, 1991, AALR, W5477, Box 873, 62/9, Part 30, Archives New Zealand, Wellington.
73. "Still a Role for Government," *Evening Post*, May 13, 1988, 6, editorial.
74. State Services Commission, Bulk Funding of Teachers Salaries: Ministry's Paper of June 7, 1991, dated June 11, 1991, AAFH, Series 6790, W5510, Box 452. SS-1-3-4-1, Part 1, Structures and Systems—Social Policy—Education, Archives New Zealand, Wellington.
75. State Services Commission, Bulk Funding of Teachers' Salaries, June 14, 1991, AAFH, Series 6790, W5510, Box 452, SS-1-3-4-1, Part 1, Structures and Systems-Social Policy—Education, Archives New Zealand, Wellington
76. *NZPD*, vol. 539, 1994, 247.
77. *NZPD*, vol. 521, 1991, 6078.
78. Ibid.
79. Ibid., 6079.
80. Ibid.
81. Ibid., 6083.
82. Ibid., 6084.
83. *NZPD*, vol. 584, 2000, 2674.
84. *NZPD*, vol. 606, 2003, 3981.
85. *NZPD*, vol. 638, 2007, 8540.
86. New Zealand Qualifications Authority, *NCEA. Better Information for Employers*, Wellington: New Zealand Qualifications Authority, fold-over pamphlet, c. 2002.
87. New Zealand Qualifications Authority, *An Introduction to NCEA*, Wellington: NCEA-001 B, c. 2002.
88. "Their Future Depends on It," *Dominion Post*, February 16, 2005, editorial, B4.
89. "NCEA Here to Stay," *Dominion Post*, February 17, 2005, A1.

90. "Must Try Harder," *Dominion Post*, February 1, 2005, A7.
91. S. Robins, "Most Secondary Teachers Support NCEA," *Otago Daily Times*, March 4, 2004, 15.

Conclusion: A Real Say or a National Morality Play? The Road to Radical Reform in Retrospect

1. *NZPD*, vol. 505, 1990, 9.
2. Ibid.,10.
3. Education Policy Response Group Occasional Paper, *Educational Reform in New Zealand 1989–1999. Is There Any Evidence for Success?* Palmerston North: Massey University College of Education, 1999, 39.
4. C. Wylie, *The Impact of Tomorrows' Schools in Primary Schools and Intermediates. 1991 Survey Report*, Wellington: NZCER, 1992.
5. Ibid., 2.
6. Ibid., 3.
7. See, for instance, "The Little Yellow School Book; The Principals of PR"; "Which Failed—System or Kids?" *New Zealand Herald*, August 29–30, 1998, H.1–3.
8. "Today's Schools. Our Findings," *New Zealand Herald*, September 2, 1998, A13.
9. "Most Parents Happy with System," ibid.
10. "Chalkface Message: Progress a Process Not an Event," ibid.
11. K. Wylie, "In Boards We Trust," *Dominion Post*, March 20, 2007, B5.
12. Ibid.
13. Ibid.
14. Ibid.
15. PPTA Executive, "Tomorrow's Schools: Yesterday's Mistake?" A paper presented to the PPTA Annual Conference, Wellington, September 30–October 2, 2008.
16. C. Wylie, "What Can New Zealand Learn from Edmonton?" Wellington: NZCER, 2007, 13, cited in "Tomorrow's Schools: Yesterday's Mistake?" 7.
17. Wylie, "Tomorrow's Schools: Yesterday's Mistake," 9.
18. Ibid., 13.
19. V. Small, "Prisons Bolster State Staff Numbers," *Dominion Post*, November 27, 2007, A2.
20. "Slice Mandarins with a Scalpel," *Dominion Post*, March 17, 2008, editorial, B4.
21. National Curriculum, Key Issues Requiring Immediate Development, c. early 1992. No author given. Private papers.
22. M. Irwin, Improving the Quality of Policy Advice—Structures and Processes, May 1992, 17–18. Private Papers.
23. J. Alison, "Mind the Gap! Policy Change in Practice. School Qualifications Reform in New Zealand, 1980–2002," PhD diss., Massey University, Palmerston North, 2007, 14.

24. H. Lee, A.-M. O'Neill, and J. McKenzie, "'To Market, To Market....' The Mirage of Certainty: An Outcomes-Based Curriculum," in *Reshaping Culture, Knowledge and Learning? Policy and Content in the New Zealand Curriculum Framework*, edited by A.-M. O'Neill, J. Clark, and R. Openshaw, Palmerston North: Dunmore Press, 2004, 47–70.
25. G. Lee, D Hill, and H. Lee, "The New Zealand Curriculum Framework: Something Old, Something New," in *Reshaping Culture, Knowledge and Learning? Policy and Content in the New Zealand Curriculum Framework*, edited by A.-M. O'Neill, J. Clark, and R. Openshaw, Palmerston North: Dunmore Press, 2004, 71–89.
26. Ibid.
27. R. Openshaw, "Citizen Who? The Debate over Economic and Political Correctness in the Social Studies Curriculum," in *New Horizons for New Zealand Social Studies*, edited by P. Benson and R. Openshaw, Massey University: ERDC Press, 1998, 20–21.
28. Ibid.
29. "Pass Rates up but 38 pc Fail NCEA Level One," *Dominion Post*, March 29, 2008, A1.
30. Ibid.
31. "Egalitarian Approach Fails Pupils," *New Zealand Herald*, 2005, A11, editorial.
32. A. Thomson, "Schools Told: Help Strugglers," *New Zealand Herald*, October 26, 2005, A1, cited in G. Lee, H. Lee, and R. Openshaw, 2007, "The Comprehensive Ideal in New Zealand: Challenges and Prospects," in *The Death of the Comprehensive High School? Historical, Contemporary, and Comparative Perspectives*, edited by B.M. Franklin and G. McCulloch. New York and Basingstoke, UK: Palgrave Macmillan, 2007, 178.
33. Ibid.
34. D. Angus and J. Mirel, *The Failed Promise of the American High School, 1890–1995*, New York: Teacher's College Press, 1999.
35. Lee, Lee, and Openshaw, "The Comprehensive Ideal," 179–180.
36. Transcript of Interview 29 with Brian Picot, November 1, 1995. G. Butterworth and S. Butterworth, Ministry of Education Oral History Project, Education Reform 1987–1995, 2.
37. Transcript of Interview 33 with Maurice Gianotti, September 10, 1996, Wellington, 8. G. Butterworth and S. Butterworth, Ministry of Education Oral History Project, Education Reform 1987–1995.
38. See, for instance, Alison, 24.
39. K. du Fresne, "Once a Cop, Always a Cop," *Dominion Post*, August 9, 2008, B5.
40. See, for instance, D. McKenzie, H. Lee, and G. Lee, *Scholars or Dollars? Selected Historical Case Studies of Opportunity Costs in New Zealand Education*, Palmerston North: Dunmore Press, 1996, 18–33.
41. See, for instance, M. Apple, *Ideology and Curriculum,* third edition, New York and London: Routledge Falmer, 2004, 4; 18–19.
42. M. Mintrom and J. Wanna, "Innovative State Strategies in the Antipodes: Enhancing the Ability of Governments to Govern in the Global Context," *Australian Journal of Political Science*, 41, no. 2, June 2006, 165–166.

43. Ibid., 174.
44. P. Chilton, "Missing Links in Mainstream CDA," in *A New Agenda in (Critical) Discourse Analysis*, edited by R. Wodak and P. Chilton, Amsterdam & Philadelphia: J. Benjamins, 2005, 19–51.
45. N. Fairclough, "Critical Discourse Analysis in Transdisciplinary Research," in *A New Agenda in (Critical) Discourse Analysis*, edited by R. Wodak and P. Chilton, Amsterdam & Philadelphia: J. Benjamins, 2005, 68.
46. T.R. Burns and M. Carson, "Social Order and Disorder. Institutions, Policy Paradigms and Discourses: An Interdisciplinary Approach," in *A New Agenda in (Critical) Discourse Analysis*, edited by R. Wodak and P. Chilton, Amsterdam & Philadelphia: J. Benjamins, 2005, 296–297.
47. H.M. Kliebard. *The Struggle for the American Curriculum 1893–1958*, third edition. New York: Routledge Falmer, 2004, 271–291.
48. See P.G. Filene. "Obituary for the 'Progressive' Movement," *American Quarterly*, vol. 22, 1970, 20–34., cited ibid., 280.
49. Daniel T. Rodgers "In search of progressivism" *Reviews in American History*, December 1982, 113–132, cited ibid., 284–285.
50. Ibid., 285–286.
51. Ibid., 282.
52. See, for instance, C. James. *The Quiet Revolution. Turbulence and Transition in Contemporary New Zealand*. Wellington: Allen and Unwin, 1986. See also C. James. *New Territory. The Transformation of New Zealand, 1984–92*. Wellington; Bridget Williams Books, 1992.
53. Kliebard, *Struggle for the American Curriculum*, 291.
54. Burns and Carson, "Social Order and Disorder," 302–303.
55. L. Webb, *The Control of Education in New Zealand*, Auckland: New Zealand Council of Educational Research, 1937, 7–8.
56. G. Butterworth and S. Butterworth, *Reforming Education. The New Zealand Experience 1984–1996*, Palmerston North: Dunmore Press, 1998, 159.
57. B. Edwards and J. Moore, "Hegemony and the Culturalist State Ideology in New Zealand," in *The Politics of Confirmity*, edited by R. Openshaw and E. Rata, Auckland: Pearson Education, 2009 forthcoming, 43.

Bibliography

Official Publications

Appendices to the Journals of the House of Representatives (AJHR)

AJHR, E-1, 2–3, 1943.
AJHR, 1946. E-1, 9, 1946.
Beeby, C.E. "A Special Topic. The Post-primary Curriculum." *AJHR*, vol. 2, E-1, 5, 1958.
Report of the Minister of Education (T.H. McCombs), 1947; *AJHR*, E-1, 5, 1948.

New Zealand Parliamentary Debates (NZPD)

NZPD, vol. 289, 1950, 211.
NZPD, vol. 291, 1950, 2529.
NZPD, vol. 469, 1985, 691.
NZPD, vol. 505, 1009, 9.
NZPD, vol. 542, 1994, 3405.
NZPD, vol. 521, 1991, 6078.
NZPD, vol. 521, 1991, 6079.
NZPD, vol. 521, 1991, 6083.
NZPD, vol. 539, 1994, 247.
NZPD, vol. 521, 1991, 6084
NZPD, vol. 584, 2000, 2674.
NZPD, vol. 606, 2003, 3981.
NZPD, vol. 638, 2007, 8540.

New Zealand Gazette

New Zealand Education Gazette, vol. 23, no. 9, 219.

Bibliography

Archival Material

Archives New Zealand (Wellington)

Department of Education and Ministry of Education Head Office Files (ABEP)

ABEP. W4262. Box 1727. 30/2/19. Part 1. English 1945–1982.
ABEP. W4262. Box 3695. NF50/2/ens-la. Part 2. Language across the Curriculum.
ASA to DA. Review of Educational Administration. July 1, 1987. ABEP, W4262, Box 923 17/5/5. Part 1. This communication was marked "urgent."
Beeby, C.E. Memo: General, April 23, 1956, re Mr Powell's pamphlet—The Maori School—a Cultural Dynamic. ABEP. W4262. E.29/2/97.
Burns, M.K. for DG, to Minister of Education, September 13, 1979. ABEP. W4262, Box 485. 1/2/3. Financial and Resource Management. Part 1.
———, August 11, 1980. ABEP. W4262. Box 485. 1/2/3. Financial and Resource Management. Part 2.
Douglas, R.O. Minister of Finance, to Russell Marshall, Minister of Education. "Review of the State Education Service" (June 9, 1987). ABEP. W4262. Box 923 17/5/5. Part 1. Review of Educational Administration.
Draft of letter sent *to New Zealand Herald*, June 19, 1956. ABEP. W4262. Box 692. Criticism. Part 1.
Education and the Economy. A Report prepared for the Organisation for Economic Co-Operation and Development (OECD) by the New Zealand Vocational Training Council. June 1986, ABEP. W4262. Box 3634. NS41/9/3. EWE.
Ensor, F. le to Beeby, C.E. June 28, 1956. ABEP, W4262. Box 692. Criticism. Part 1.
Financial Management and Review Team. Review of Financial Management and Control. The Treasury: Wellington, April 1978. ABEP. W4262. Box 485. 1/2/3. Financial and Resource Management. Part 1.
Hauiti Marae. Programme Draft, April 1977. ABEP. W4262. Box 2677. 40/1/4/5. Marae-based courses for teachers. Part 1.
Head Office Submissions to National Advisory Committee on Maori Education, c. 1976. ABEP. W4262. Box 2848. 53/2/39. Maori. Part 3.
Hunter, J.L. Chairman, National Advisory Committee. Memo entitled, "Representation on the National Advisory Committee on Maori Education." Undated but c. 1969. W4262. Box 2898. 46/3/14. National Advisory Committee on Maori Education.
Irvine, H. for DG. Conference: "Education and the Equality of the Sexes," c. 1975. ABEP. W4262. Box 1525. 26/1/174/1. Part 1. Education and the Equality of the Sexes.
Koopman-Boyden to Marshall, September 28, 1984. ABEP. W4262. Box 1728. 34/2/17. Report on the XXth International Committee on Family Research

Seminar on Social Change and Family Policies, Melbourne, August 19–24, 1984. Social Education. Part 4.
le Ensor, L.F. to Beeby, C.E., June 28, 1956. ABEP. W4262. Box 692. Criticism. Part 1.
Letter from participant to the organisers of the Te Wai Pounamu College Course, Dunedin, October 31, 1977. ABEP. W4162. Box 2677. 40/1/4/5. Marae-based Courses for Teachers. Part 1.
Marae based courses, Mangamuka, Hokianga, May 29–June 2, 1977. Handout. ABEP. W4262. Box 2677. 40/1/4/5. Marae-based courses for teachers. Part 1.
Miller, C.H.S. for DG, to Associate Minister of Finance, September 20, 1978. ABEP. W4262. Box 485. 1/2/3. Financial and Resource Management. Part 3.
Miller, C.H.S. for DG to Secretary of the Treasury, November 9, 1978. ABEP. W4262. Box 482. 1/2/2. Miscellaneous Correspondence.
Minister of Education to G. Powell, February 28, 1956. Education Systems—Maori Education 1949–56. ABEP. W4262. E29/2/97.
National Advisory Committee on Maori Education. "Priorities for Developments." July 1970. ABEP. W4262. Box 2848. 53/2/39. Maori/ Part 2.
Notes of Minister's meeting with Education Board Chairmen held on February 4, 1982. ABEP. W4262. Box 485. 1/2/8. Budget Preparation. Part 9.
Palmerston North Teachers College. Te Kooti Project, 1978. Project proposal dated December 4. ABEP. W4262. Box 3066. 52/2/53. Life of Te Kooti.
Quigley, D.F. to Chairman, Cabinet Committee on Expenditure, March 14, 1980. ABEP. W4262, Box 485. 1/2/3. Financial and Resource Management. Part 2.
Renwick, W.L. to Whitehead, C., April 2, 1973, 2.ABEP. W4262. Box 1704. 34/1/23. Post primary Curriculum (Thomas Report).
Renwick, W.L. for DG to Minister of Education, Review of secondary curriculum, April 16, 1973. ABEP. W4262. Box 1836. Box 1836. 34/2/13. Part 2. Review of Secondary Curricula.
Renwick, W.L. to Minister, January 6, 1977. Women and Education—Policy and Actions. ABEP. W4262. Box 1525. 26/1/174/1. Part 1. Education and the Equality of the Sexes.
Renwick, W.L. to Minister of Education, July 25, 1978. ABEP. W4262. Box 485. 1/2/3. Financial and Resource Management. Part 1.
———, Review of Secondary Curriculum, April 16, 1973. ABEP. W4262. Box1836. 34/2/13. Part 2. Review of Secondary Curricula.
Renwick, W.L. Comments on the Treasury Report re-the Development of the Psychological Service dated March 28, 1979. ABEP. W4262, Box 485. 1/2/3. Financial and Resource Management. Part 3.
Renwick, W.L. to Minister of Education, May 19, 1980. ABEP. W4262, Box 485. 1/2/3. Financial and Resource Management. Part 2.
Renwick, W.L. A reconsideration of the core curriculum in primary and secondary schools, c. August 1983. ABEP. W4262. Box 1836. 34.2.13. Part 3. Review of Secondary Curricula.
Renwick, W.L. to Chairman, Waitangi Tribunal, May 28, 1986. ABEP. W4262. 5/1/12. Treaty of Waitangi. Part 1.

Renwick, W.L. Departmental minute 1987/20. To all staff from Head Office, June 12, 1987. ABEP, W4262. Box 923.17/5/5. Part 1. 17/2/16 NS.

The Report of the Controller and Auditor General on Financial Management in Administrative Government Departments. The Treasury: Wellington, May 1978. ABEP. W4262. Box 485. 1/2/3. Financial and Resource Management. Part 1.

Report on the Primary National Course Lopdell House, Teachers Colleges, August 30–September 3, 1971. Report entitled "Teachers College Courses which prepare teachers to provide for the needs of Maori and Island children." ABEP. W4262. Box 2898. 46/3/14. National Advisory Committee on Maori Education.

Secretary to the Treasury. Task Force Report: New machinery for public sector planning. The Treasury: Wellington. August 1977. ABEP. W4262, Box 485. 1/2/3. Financial and Resource Management. Part 2.

Secretary of Treasury to Education Department Accountant, September 17, 1979. ABEP. W4262, Box 485. 1/2/3. Financial and Resource Management. Part 2.

Sheen, K. to Minister of Education, August 12, 1970. ABEP. W4262. Box 692. 17/1/25. Criticism. Part 1.

Simpson, M. "The Handbook. The Project," c. July 1967. ABEP. Box 2848. 53/2/39. Maori. Part 1.

Smith, A.F. Letter to Bernie Kernot, November 15. ABEP. W4262. Box 2677. 40/1/4/5. Marae-based courses for teachers. Part 1.

Smith, A.F. National Advisory Committee on Maori Education. Report of the Office for Maori and Island Education, 1971. ABEP. W4262. Box 2898. 46/3/14. National Advisory Committee on Maori Education.

Submission from Citizens Association for Racial Equality, December 10, 1984. ABEP. W4262. Box 1830, 31/1/64/4. Part 1. Inquiry into Curriculum: Assessment and Qualifications.

Sutton, N.F. to DG. Task Force Report: Re-New Machinery for Public Sector Planning. August 16, 1977. APEP. W4242. Box 485. 1/2/3. Financial and Resource Management. Part 2.

Sutton, N.F. for DG to Secretary for the Treasury, February 21, 1980. ABEP. W4262, Box 485. 1/2/3. Financial and Resource Management. Part 1.

Taha Maori across the Curriculum, 1983. No author listed. ABEP. W4262. Box 3732. NS50/2/MA-TM. Maori across the Curriculum.

The Treasury, Task Force Report: New Machinery for Public Sector Planning. The Treasury: Wellington, August 10, 1977. ABEP. W4242. Box 485. 1/2/3. Financial and Resource Management. Part 2.

Unsigned, handwritten memo dated July 7, 1987. ABEP, W4262, Box 923 17/5/5. Part 1.

University of Waikato. Centre for Maori Language Studies and Research. *Heads Held High.* Te Kohanga Preschool Project Report for 1974. Working Paper No. 5. December. ABEP. W4262. Box 2848. 53/2/39. Maori. Part 3.

Vocational Guidance and Education for Work: Working Party Recommendations. ABEP. W4262. Box 1525: 26/1/174/1: Part 1. Education and the Equality of the Sexes.

Wanganui Office of Te Puni Kokiri (ABRP)

Anna Jones, for Secretary. Dept of Maori Affairs, March 29, 1983, to District Officers. ABRP. 6844. W4598. Box 200. 31/3/1. Part 6. Te Kohanga Reo-Programme.

Report on Te Kohanga Reo Operations nationally, dated September 1983, entitled: Te Kohanga Reo (The Language Nest). ABRP. 6844. W4598. Box 200. 31/3/1. Part 6. Te Kohanga Reo-Programme.

Ross for DG to District Senior Inspectors of Primary Schools, August 1, 1983. ABRP. 6844. W4598. Box 200. 31/3/1. Part 6. Te Kohanga Reo-Programme.

State Services Commission Files (AAFH)

Birch, W.F. Memorandum for Cabinet Expenditure Control Committee entitled, "Today's Schools- Report on the Ministry of Education's Outputs," March 1, 1991. AAFH. Series 6790. W5510. Box 458. SS-1–3–4. Part 1. Structures and Systems—Social Policy- Education—Schools—Policy.

Cabinet Expenditure Control Committee. "Education Expenditure." March 15, 1991. AAFH. Series 6790. W5510. Box 458. SS-1–3–4. Part 1. Structures and Systems—Social Policy—Education—Schools—Policy.

Cabinet Expenditure Control Committee. Minutes of a Meeting held on March 12, 1991. AAFH. Series 6790. W5510. Box 458. SS-1–3–4. Part 1. Structures and Systems—Social Policy—Education—Schools—Policy.

Carpinter, P. "Principles Underlying the State Sector; the SSD View." September 20, 1990 (12pp). AAFH. Series 6790. W5510. Box 438. SS-1–3–1-X. Part 1. General Policy.

Comments c. March 1991. AAFH. Series 6790. W5510. Box 452. SS-1–3–4–1. Part 1. Structures and systems-social policy-education.

Extracts from Conversation between Graye Shattky (G) and Lockwood Smith (L)—Kou Lounge, Auckland, March 2, 1991, 6pm to 7.25pm. AAFH. Series 6790. W5510. Box 452. SS-1–3–4–1. Part 1. Structures and Systems-Social Policy-Education.

Gibson, C. Group Manager Operations, for Review Team (Ministry of Education), to Treasury, March 6, 1991. AAFH. Series 6790. W5510. Box 458. SS-1–3–4. Part 1. Structures and Systems—Social Policy—Education—Schools—Policy.

Gill, J. to Greig, D., May 24, 1991. AAFH. Series 6790. W5510. Box 458, SS-1–3–4. Part 1. Structures and Systems—Social Policy—Education—Schools—Policy.

Greig, D., for Secretary to the Treasury to Minister of Finance. Memo entitled, "Ministry of Education: Size of Outputs." March 4, 1991 (4 pp.). AAFH. Series 6790. W5510. Box 458. SS-1–3–4. Part 1. Structures and Systems—Social Policy—Education—Schools—Policy.

Greig, D. to D. Tripp, May 2, 1991. AAFH. Series 6790. W5510. Box 458. SS-1–3–4. Part 1. Structures and Systems—Social policy—Education—Schools—Policy.

Hunn, D.K. "Overview of Key Issues." October 29, 1990. AAFH. Series 6790. W5510. Box 438. SS-1-3-1-X. Part 1. General Policy.

Inter Office Memorandum from Tripp, D. Hitchiner S. and Greig, D. entitled "Ministry Outputs," March 13, 1991. AAFH. Series 6790. W5510. Box 458. SS-1-3-4. Part 1. Structures and Systems—Social Policy—Education—Schools—Policy.

Main Points from School Visits, c. March 1991. AAFH. Series 6790. W5510. Box 452. SS-1-3-4-1. Part 1. Structures and Systems-Social Policy-Education.

The Major Issues for Education, Science and Technology Official's Committee. February 4, 1991. AAFH. Series 6790. W5510. Box 458. SS-1-3-4. Part 1. Structures and Systems—Social Policy- Education—Schools—Policy.

Pilgrim, C. STA Special Report Rural Schools. The Fabric of the Rural Community. March 1991. AAFH. Series 6790. W5510. Box 452. SS -1- 3–4 -1, pt 1. Structures and Systems-Social Policy-Education.

Review Committee Report, Learning Media, dated April 10, 1991. AAFH. 6790. W5510. Box 458, SS-1-3-4. Part 1. Structures and Systems—Social Policy—Education—Schools—Policy.

Smith, L.M. A Strategy for Investment in Education and Training. Draft 3. May 1991. Addressed to Chair, Cabinet Strategy Committee, Executive Summary, AAFH. Series 6790. W 5510. Box 438. SS-1-3-1-X. Part 1. General Policy.

State Services Commission. Bulk Funding of Teachers' Salaries. June 14, 1991. AAFH. Series 6790. W5510. Box 452. SS-1-3-4-1, Part 1. Structures and Systems—Social Policy—Education.

———. Bulk Funding of Teachers Salaries: Ministry's Paper of June 7, 1991, dated June 11, 1991. AAFH. Series 6790. W5510. Box 452. SS-1-3-4-1. Part 1. Structures and Systems—Social Policy—Education.

———. Learning Media Review, March 20, 1991. AAFH. Series 6790. W5510. Box 458. SS-1-3-4. Part 1. Structures and Systems—Social Policy—Education—Schools—Policy.

State Services Commission Surveys. Boards of Trustees, c. March 1991. AAFH. Series 6790. W5510. Box 452. SS-1-3-4-1. Part 1. Structures and Systems-Social Policy-Education.

Treasury Files (AALR)

Anderton, J. "Output-based" Advice from The Treasury Factory to Sell Schools. Press Release, May 13, 1991. AALR. W5477. Box 873. 62/9. Part 30.

C.E.I.S. Evaluation Report. Otara Resource Network. Interim Report. September 11, 1984. AALR. Series 873. W444. Box 247. 52/479/1. Part 2.

Community Education Initiative Scheme: Meeting with Ministers, Parliament Buildings, October 10, 1984 12.00–1.00pm, 2. AALR. Series 873. W444. Box 247. 52/479/1. Part 2.

Document entitled T89/140. September 13, 1989. Appended to Goff, P. Memorandum to Cabinet Social Equity Committee. Ministry of Education. AALR 873. W5427. Box 1065. 62/9. Part 29. Education—Miscellaneous.

Goff, P. Memorandum to Cabinet Social Equity Committee. Ministry of Education. Undated but c. late 1989. AALR. 873. W5427. Box 1065. 62/9. Part 29. Education—Miscellaneous.

Lange, Rt Hon. (David). Education Media Briefing. April 13, 1988. AALR, 873, W5427/164. 62/9. Part 28.

———. Memo to the New Zealand Secondary School Boards Association Biennial Conference. Central Institute of Technology, Trentham. Speech notes. August 28, 1987. AALR. Series 73. W5427. Box 1064. 62/9. Part 27. Education Miscellaneous.

Lange, Rt Hon. (David) to PPTA Conference. Sheraton Hotel, Auckland, August 25, 1987. Speech notes. AALR. Series 73. W5427. Box 1064. 62/9. Part 27, Education Miscellaneous.

Minister of Finance. October 8, 1984. Community Education Initiatives Scheme. AALR. Series 873. W444. Box 247. 52/479/1. Part 2.

Minutes of the CEIS Interdepartmental Committee Meeting held on Thursday, September 20, 1984. AALR 873. W444. Box 247. 52/479/1. Part 2.

Report of the Committee on Gangs. Government Response. AALR. Series 873. W444. Box 247. 52/479/1. Part 2.

David Lange Papers (AAWW)

Administration of Education. Underlying Efficiency Considerations. Outline note by Treasury. AAWW. Series 7112. Box 9. Education Including Picot Taskforce.

Briefing Notes for Press Conference, August 7, 1988. AAWW. Series 7112. W4240. Box 19. Teacher HR, Picot Implementation.

Confidential Letter from Logan, S. Managing Director, to McQueen, H, PMs Department, June 8, 1988. AAWW. Series 7112. W4640. Box 4: Logos/Picot, Curriculum Review.

Draft Submission to Taskforce on Education Administration (Department of Education). Undated but c. October 1987. AAWW. Box 9. Education Including Picot Taskforce.

Knight, G. to Ross, J.A. Picot Report and the Integration Act, July 3, 1988. AAWW. 7112. W4640. Box 24. Social Equity.

Lange, the Rt Hon (David) to Chair of Cabinet Social Equity Committee. Picot Taskforce: Officials Committee, April 11, 1988. AAWW. Series 7112. W4640. Box 24. Social Equity.

Lange, M. to Lange, the Rt Hon. (David), March 21, 1989. AAWW. Series 7112. Box 9. Education Including Picot Taskforce.

LOGOS. Letter Analysis, June 16, 1988, AAWW. Series 7112. W4640. Box 24. Social Equity.

McQueen, H. Memo to John Henderson. March 2, 1988. AAWW. Series 7112. W4640. Box 9. Education Including Picot Taskforce.

———. The Year of Educational Change: A Strategy to Implement the Picot Report and Enhance the Government's Image in Education. C. February 1988. AAWW. Series 7112. W4640. Box 9. Education Including Picot Taskforce.

Minister of Education Releases Picot Report. Press statement dated May 10, 1988. AAWW. Series 7112. W4640. Box 4: Logos/Picot, Curriculum Review.

National Survey—Key findings and recommendations, dated July 26, 1988. Insight New Zealand. AAWW. Series 7112. W4640, Box 4: Logos/Picot, Curriculum Review.

Outline Approach to Implementation Phase June 7, 1988. AAWW. Series 7112. W4640. Box 4: Logos/Picot, Curriculum Review.

Picot Implementation Phase Budgets and General Strategy and Submission Analysis. AAWW. Series 7112. W4640. Box 4: Logos/Picot, Curriculum Review.

The Picot Report Survey. Detailed Tabular Report. AAWW. Series 7112. W4640. Box 4: Logos/Picot, Curriculum Review.

Picot Video Script. AAWW, 7112, W4262, Box 5. National Library, Picot.

Report of the Officials Committee on the Reform of Educational Administration: Revised Paper, July 25, 1988. AAWW. Series 7112.W4640. Box 24. Social Equity.

Responses to the Picot Report, July 13, 1988. AAWW. 7112. W4640. Box 24. Social Equity.

Ross, A. Acting DG, to Logan, S., Managing Director, LOGOS Public Relations Ltd, April 29, 1988, AAWW. Series 7112. W4640. Box 4: Logos/Picot, Curriculum Review.

Smelt, S.J., Accountability in Education. Note by Secretariat to Taskforce to Review Education Administration. November 9, 1987. AAWW. Box 9. Education including Picot Taskforce.

Submission from Nga Iwi Maori—from the Maori People throughout New Zealand. Tabled by Wiremu Kaa. July 1, 1988. AAWW. Series 7112. W4640. Box 24. Social Equity.

Taskforce to Review Education Administration, Possible Organisational Models. Note by Treasury (Undated, no author supplied), AAWW. Box 9. Education Including Picot Taskforce.

Department of Education Head Office.
Residual Management Unit (AAZY)

Bazley, M. The Watts Report on New Zealand Universities and Post-compulsory Education and Training. Social Services Division, SSC. Wellington, February 22, 1988. AAZY. W3901. Box 181. Picot-General.

Dixon, A., to PM, February 15, 1988, Picot Taskforce—Progress Report. AAZY. W3901. Box 181. Picot-General.

Folder 15. Item 79. Untitled thirty page document providing abstracts of readings for the Picot Taskforce. AAZY. W.3901. Box 181. Picot-General.

Gianotti, M. Notes of a meeting of the first meeting of the Taskforce held in Wellington July 31, 1987. AAZY W3901. Box 181. Picot–General.

Government Research Unit. Education Vouchers. February 4, 1987. Folder 15. AAZY. W3901. Box 181. Picot-General.
Macpherson, R.J.S. Report on Structural Changes in Victoria's Education, SSC, March 2, 1988. AAZY. W3901. Box 181. Picot-General.
Taskforce to Review Educational Administration. Meeting 8.00am to 4.30pm Friday July 31, 1987. AAZY. W3901. Box 181. Picot–General.
Task Force to Review Education Administration. Record of Policy Points Made and Tasks to be Undertaken from Meeting Number Two of August 17, 1987. AAZY.W3901. Box 181. Picot–General.
Task Force Status Report 31. 8. 87. AAZY.W3901. Box 181. Picot–General.
Untitled 30 page-document. Folder 15. AAZY. W3901. Box 181.Picot-General.

Currie Report—Submissions to

Associated Chambers of Commerce of New Zealand, Submission. E.50/13.
Blake, N., Effect of Educational Administration on Emotional Instability in Society. Submission. E50/8.
Murdoch, J.H. Submission. E50/12.
New Zealand Department of Education. Submission. No. 11A. E50/8.
New Zealand Engineering and Related Industries Association and Metal Trades Employers Association of Wellington (Inc), Submissions. E50/14.
Opotiki College Board. Submission. E50/8.
Sutch, W.B., Submission. E.50/15.
Timaru School Board. Submission. E50/8.

Archives New Zealand (Auckland)

Department of Education. Northern Regional Office (BAAA)

Green, S.S. The Case for a Commission of Inquiry or Consultative Committee to Investigate the Administration of Education in New Zealand. Unpublished paper, November 18, 1954. BAAA. A841/30.

Regional Department of Education Residual Management Unit Files (BCDQ)

Atkinson, P.A. to DG, Assistant DGs, Assistant Secretaries, National Superintendents of Education, and all Directors. Entitled 'Revision of the Education Act—Progress Report, July 17, 1986, BCDQ. 4031D. A739. 1/15. Revision of Education Act.
Holmes, F. Lay and Professional Participation in Educational Administration. Unpublished paper. C. early 1976. BCDQ. 5193R. A739. Lay and Professional Participation in Educational Administration, 1976.

Memo entitled, "Ministry of Education Setting Up" (as at November 21, 1988). BCDQ. A739 4227a. 1/17/1. New Ministry of Education.
Moody, P.T. for DG, Head Office Minute 1975/42. Reorganisation of Education Administration. Dated October 15, 1975. BCDQ. 5193N. A739. Reorganisation of Educational Administration, 1976.
New Release from the State Services Commission. July 4, 1989. BCDQ. A739. 4227a. 1/17/1. New Ministry of Education.

Auckland Education Residual Management Unit (YCDB)

"CRIS Curriculum Review in Science: Discussion Paper 8." Multicultural Science, c. 1986. YCDB. A688. 152b. 33/7. Secondary Curriculum. Part 1.
Curriculum Development Division, Department of Education. Hui Whakamatatau, Taurua Marae, Rotoiti, November 7–9, 1984. 1. Auckland Education Board Residual Management Unit. YCDB. A688. 1591a. 33/1. Maori Studies. Part 2. 1985–1987.
Taha Maori and Change. Report from National Residential Inservice Course, Lopdell Centre, July 14–18, 1986. YCDB. A688. 1591a. 33/1. Maori Studies. Part 1, 1985–1987.
Te Kete Timatatanga. "A Teacher Guide to Taha Maori in Social Studies." Booklet enclosed with a letter from Ash Newth, District Senior Inspector of Primary Schools to nga Rangatira (Principals), October 2, 1985. YCDB. A688. 1591a. 33/1. Maori Studies. Part 2, 1985–1987.

Private Papers (Held by Author)

National Curriculum. Key Issues Requiring Immediate Development. C. early 1992. No author supplied.
Irwin, M. Improving the Quality of Policy Advice—Structures and Processes. May 1992.

Newspapers

Auckland Star (Auckland Daily)

"Care Asks for Commission on Maori Schooling." *The Auckland Star*, April 14, 1984, A3.
Glasgow, L. "Shift in Education Needed," in "Today's Woman" Feature, *Auckland Star*, March 28, 1984, B3.

Dominion (Wellington Daily)

Note: This newspaper was known as the *Dominion* until amalgamation with the *Evening Post* in July 2002. Thereafter it became the *Dominion Evening Post*.

Dawson, D. "Education Response of Concern to Lange." *Dominion*, August 26, 1987, 7.
"Education Reshuffle." *Dominion*, June 24, 1987, 14, Editorial.
du Fresne, K. "Once a Cop, Always a Cop." *Dominion Post*, August 9, 2008, B5.
Lagan, B. "Spending Cuts Target Higher." *Dominion*, March 5, 1982, 1.
"NCEA Here to Stay." *Dominion Post*, February 17, 2005, A1.
"Pass Rates up but 38 pc Fail NCEA Level One." *Dominion Post*, March 29, 2008, A1.
Political Reporter, "Where Your Money Should NOT Have Gone." *Dominion*, June 28, 1978, 1.
"Slice Mandarins with a Scalpel." *Dominion Post*, March 17, 2008, B4, Editorial.
Small, V. "Prisons Bolster State Staff Numbers." *Dominion Post*, November 27, 2007, A2.
"Teacher Says Cuts Add to Rejects." *Dominion*, March 9, 1982, 3.
"Their Future Depends on It." *Dominion Post*, February 16, 2005, B4, Editorial.
Wylie, C. "In Boards We Trust." *Dominion*, March 20, 2007, B5.

Evening Post (Wellington Daily)

Note: This newspaper was known as the *Evening Post* until amalgamation with the *Dominion* in July 2002. Thereafter it became the *Dominion Evening Post*.

"'Aberration' Needs Study." *Evening Post*, April 11, 1987, 68.
Brown, K. Education Reporter. "Post Exclusive; Sell Off Schools, Treasury Suggests." *Evening Post*, May 10, 1991, 1.
"Cut Condemned by Inspectors." *Evening Post*, March 9, 1982, 27.
"Expenditure Cuts Could Reach 8 1/2 pc." *Evening Post*, March 5, 1982, Section 1, 4.
"National Offers Support." *Evening Post*, May 11, 1988, 2.
"Now for a Lesson in Education." *Evening Post*, May 11, 1988, 6, Editorial.
"PM Wants to Axe Dept of Education." Front page lead. *Evening Post*, November 7, 1987, 1.
"Picot Report Member Changes His Mind." *Evening Post*, July 25, 1988, 1.
"Still a Role for Government." *Evening Post*, May 13, 1988, 6, Editorial.

Evening Standard (Palmerston North Daily)

Note: This newspaper was known as the *Evening Standard* until July 2002. Thereafter it became the *Manawatu Standard*.

"Kids Back in Class But Dispute Drags On." *Evening Standard*, February 24, 1978, 1.

New Zealand Herald (Auckland Daily)

Corbett, J., du Chateau, C., Burge, K., Boland, M.-J., and Knight, R. "The Little Yellow Schoolbook." *New Zealand Herald*, August 29–30, 1998, H1–3.
"Chalkface Message: Progress a Process Not an Event." *New Zealand Herald*, September 2, 1998, A13.
"Egalitarian Approach Fails Pupils." *New Zealand Herald*, 2005, A11, Editorial
"Most Parents Happy with System." *New Zealand Herald*, September 2, 1998, A13.
"Professor Reveals Pressure Put on the Picot Report." *New Zealand Herald*, July 23, 1988, 5.
"The Principals of PR." *New Zealand Herald*, August 29–30, 1998, H2
Thomson, A. "Schools Told: Help Strugglers." *New Zealand Herald*, October 26, 2005, A1.
"Today's Schools. Our Findings." *New Zealand Herald*. September 2, 1998, A13.
"Trim PM Tackles 'New Right.'" *New Zealand Herald*, July 25, 1988, 9.
"Which Failed—System or Kids?" *New Zealand Herald*, August 29–30, 1998, H3.

Northern Advocate (Whangarei, Daily)

"Picot Report Backed." *Northern Advocate*, May 11, 1988, 2, Editorial.

The Press (Christchurch Daily)

"The Picot Report." *The Press*, May 12, 1988, 12, Editorial.

Star (Christchurch Daily)

Note: This newspaper was known as the *Christchurch Star* from 1930–1980. Thereafter it became the *Star*.
"Charging of Subject Fees Slated by PPTA." *Christchurch Star,* August 24, 1977, 22.
"Discrimination Say Teachers." *Christchurch Star,* August 23, 1977, 82.
"Education Will Never Be the Same Again." *The Star*, May 11, 1988, 8. Editorial.
Long, J., Education Reporter. "Tighter Control for Schools over Funds?" *The Press*, June 9, 1988, 8.

Staff Reporter. "Education Spending: Is Record Enough?" *Christchurch Star,* August 23, 1977, 1.
"Teachers Protest over Salary Claims." *Christchurch Star,* June 21, 1978, 24.

Waikato Times (Hamilton Daily)

Hubbard, G. "'New Right' Threat to Teacher Unions." *Waikato Times,* May 18, 1988, 3.

Interview Transcripts: Butterworth, G. and Butterworth, S. Ministry of Education Oral History Project. Education Reform 1987–1995

Interview 5 with Dr Simon Smelt. Chief Analyst with Treasury, formerly Manager Education Section. March 23, 1994. Wellington.
Interview 20 with Whetu Wereta. Picot Taskforce Member. June 20, 1994. Wellington.
Interview 29 with Brian Picot. Chairman. Picot Taskforce. November 1, 1995. Auckland.
Interview 33 with Maurice Gianotti. Chief Executive to Picot Taskforce. September 10, 1996. Wellington.

Radio New Zealand Transcripts (Wellington)

Newzel Log. Radio New Zealand National Programme, May 14, 1988. Transcript.

Published Reports and Government Agency Publications

Commission of Inquiry into an Alleged Breach of Confidentiality of the Police File on the Honourable Colin James Moyle, M.P. Wellington: Prime Minister's Department Press Office, 1978.

Fargher, R. and Probine, M. *The Report of a Ministerial Working Party. The Management, Funding and Organisation of Continuing Education and Training.* Wellington, March, 1987.

The Crewe Enquiry: Investigation into Certain Aspects of the Evidence Relating to the Conviction for Murder of Arthur Allan Thomas. A Second Confidential Report by R.A. Adams-Smith Q.C. for the Prime Minister, the Right Honourable R.D. Muldoon. Wellington, December 1979.

Education and Science Select Committee (1986). *Report of the Inquiry into the Quality of Teaching. Second Session, Forty-first Parliament* (Noel Scott, Chairman). Wellington: Government Printer.

Foreign Affairs and Defence Select Committee. Report of the Inquiry into Disarmament and Arms Control, 1985, Wellington: Government Printer.

Inquiry into Reported Allegations of Police Misuse of the Wanganui Computer. Report of the Wanganui Computer Centre Privacy Commissioner. Wellington: Government Printer, May 1986.

New Zealand Department of Education. *OECD Review of New Zealand Educational Policies 1982. Draft Statement for the Examining Panel.* New Zealand Wellington: Department of Education, February 1982.

———. *Three Years On. The New Zealand Education Reforms.* Wellington: Learning Media, 1993.

New Zealand Qualifications Authority. *Learning to Learn. An Introduction to the New Zealand Qualifications Framework.* Wellington: New Zealand Qualifications Authority, 1992.

———. *NCEA. Better Information for Employers.* Wellington: New Zealand Qualifications Authority, Fold-over pamphlet, c. 2002.

———. *An Introduction to NCEA.* Wellington: New Zealand Qualifications Authority, c. 2002. NCEA-001 B.

New Zealand Vocational Training Council. *Education and the Economy.* A Report prepared for the Organisation for Economic Co-Operation and Development. June 1986.

OECD. *Review of National Policies for Education: New Zealand.* Paris, 1983.

Picot, B. *Working Together,* Wellington: NZ Planning Council, June 1978. NZPC No. 7.

Report of the Advisory Council on Educational Planning (The Holmes Report), Wellington: Government Printer, 1974.

Report of the Commission on Education in New Zealand (The Currie Report). Wellington: Government Printer, 1962.

Report of the Royal Commission to Inquire into the Circumstances of the Conviction of Arthur Allan Thomas for the Murders of David Harvey Crewe and Jeanette Lenore Crewe, Wellington, 1980.

The Report of the Committee of Inquiry into Allegations concerning the Treatment of Cervical Cancer at National Women's Hospital and into Other Related Matters, Auckland: Government Printer, July 1988.

Report of the Committee on the Post-Primary School Curriculum (The Thomas Committee). Wellington: Department of Education, 1944 (1959 reprint).

Report of the Committee on New Zealand Universities (The Parry Report). Wellington: Department of Education, 1959.
Report on Department of Maori Affairs (The Hunn Report), Wellington: Government Printer, 1960.
Report of the Royal Commission to Inquire into the Crash of a DC10 Aircraft operated by Air New Zealand Limited. Presented to the House of Reps by Command of His Excellency the Governor-General (Sir Keith Holyoake). Wellington: Government Printer, 1981.
Report of the Taskforce to Review Educational Administration. *Administering for Excellence. Effective Administration in Education* (The Picot Report). Wellington. Government Printer, 1988.
The Royal Commission on Social Policy. *How Fair Is New Zealand Education? Part 1 How Fair Is New Zealand?*: NZCER. November 1987.
———. *How Fair Is New Zealand Education? Part 11. Fairness in Maori Education. A Review of Research and Information.* NZCER. November. (Edited by R. Benton). *Tomorrows' Schools. The Reform of Education Administration in New Zealand.* Wellington: Government Printer, August 1988.
Sexton, S. *New Zealand Schools: An Evaluation of Recent Reforms and Future Directions.* Wellington: New Zealand Business Roundtable. Wellington, 1990.
Tauroa, H. *Race against Time.* Wellington: Human Rights Commission, 1982.
Today's Schools. A Review of the Education Reform Implementation Process (The Lough Report). Prepared for the Minister of Education, Wellington, April, 1990.
The Treasury. *Government Management. Brief to the Incoming Government 1987.* Vol. 1 (untitled). Wellington: Government Printer, August 1987.
———. *Government Management. Brief to the Incoming Government 1987.* Vol. 11. Education Issues. Wellington: Government Printer, August 1987.
Vocational Training Council. Education and the Economy. A Report prepared for the Organisation for Economic Co-Operation and Development by the New Zealand Vocational Training Council. June 1986.
W.D. Scott Deliotte Ltd. *Report to the Taskforce to Review Education Administration. Initial Assessment of the Functions of the Department of Education.* November 1987.
White Paper. *The National Qualifications Framework of the Future.* Wellington: Ministry of Education, 1999.

Books and Articles

Abercrombie, M., Hill, S., and Turner, B.S. *Sovereign Individuals of Capitalism.* London, Boston: Allen and Unwin, 1986.
Alcorn, N. *"To the Fullest Extent of His Powers." C.E. Beeby's Life in Education.* Wellington: Victoria University Press, 1999.

Alison, J. "Mind the Gap! Policy Change in Practice. School Qualifications Reform in New Zealand, 1980–2002." PhD diss. Massey University, Palmerston North, 2007.

Anderson, W. *The Flight from Reason in New Zealand Education*. Auckland: Catholic Teachers' Association of Auckland, 1944.

Andrews, C. "An Analysis of Some of the Major Changes in Girls' Secondary Education since World War Two." MA diss. University of Auckland, 1980.

Angus, D. and Mirel, J. *The Failed Promise of the American High School, 1890–1995*. New York: Teacher's College Press, 1999.

Apple, M. *Ideology and Curriculum*. Third edition. New York and London: Routledge Falmer, 2004.

———. "Ideology, Equality and the New Right." In *Education Policy and the Changing Role of the State*. Proceedings of the New Zealand Association for Research on Education Seminar on Educational Policy, *Delta Studies in Education*, no. 1, edited by L. Gordon and J. Codd. Massey University, Palmerston North, July 1990.

Barcan, A. *Sociological Theory and Educational Reality. Education and Society in Australia since 1949*. Kensington, NSW University Press, 1993.

Barrington, J. "'Learning the Dignity of Labour': Secondary Education Policy for Maoris." *New Zealand Journal of Educational Studies*, 23, no. 1, 1988, 45–58.

———. "The Politics of School Government." In *The Politics of Education in New Zealand*, edited by M. Clark, 66–88. Wellington: NZCER, 1981.

———. "Historical Factors for Change in Education." In *Redistribution of Power? Devolution in New Zealand*, edited by P. McLinlay, 191–213. Wellington: Victoria University Press for Institute of Policy Studies, 1990.

———. "Why Picot? A Critique and Commentary." *New Zealand Journal of Educational Administration*, 5, November 1990, 15–17.

Barry, B. *Culture and Equality. An Egalitarian Critique of Multiculturalism*. Polity Press, Blackwell: Oxford, 2001.

Bates, R.J. "The New Sociology of Education: Directions for Theory and Research." *New Zealand Journal of Educational Studies*, 13, no. 1, May 1978, 3–14.

Beeby, C.E. *The Biography of an Idea. Beeby on Education*. Educational Research Series No. 69. Wellington: NZCER, 1992.

Bell, A. "Cultural Vandalism and Pakeha Politics of Guilt and Responsibility." In *Tangata Tangata. The Changing Ethnic Contours of New Zealand*, edited by P. Spoonley, C. McPherson and D. Pearson, 89–107, Southbank, Victoria, Australia: Thomson/Dunmore Press, 2004.

Benton, R. *How fair Is New Zealand Education? Part 11. Fairness in Maori Education. A Review of Research and Information*. The Royal Commission on Social Policy: Wellington: NZCER, 1987. November.

Birks, S. and Chatterjee, S. *The New Zealand Economy. Issues and Policies*. Second edition. Palmerston North: Dunmore Press, 1992.

Blaiklock, E.M. "Whistling Out of Fashion," *New Zealand Herald*, June 27, 1978. In *The Best of Grammatricus. Writings of Professor E.M. Blaiklock*, edited by D. More, 54, Auckland: Wilson and Horton, 1984.

Blaiklock, E.M. *Between the Morning and the Afternoon*. Palmerston North. Dunmore Press, 1980.

Boston, J., Haig, B., and Lauder, H. "The Third Wave: A Critique of the New Zealand Treasury's Report on Education. Part 11." *New Zealand Journal of Educational Studies*, 23, no. 2, 1988, 115–144.

Bowler, J. "The New Zealand Controversy over the Johnson Report: The Context of the Report of the Committee on Health and Social Education," *Growing, Sharing, Learning* (1977). PhD diss., Massey University: Albany, New Zealand, 2004.

Brame, J. "Towards Supermarket Education." *Broadsheet*, 160, July/August 1988, 6–7; 9–10.

Bray D.H. "Attitudes and Values of Polynesian and Pakeha: Social and Educational Implications of Research Findings." In *Contemporary New Zealand. Essays on the Human Resource, Urban Growth and Problems of Society*, edited by K.W. Thomson and A.D. Trlin, 174–184. Hong Kong: Hicks Smith and Sons in association with the Geography Department, Massey University, Palmerston North, 1973.

Broadsheet, no. 20, July 1974, 2, Editorial.

Butterworth, G. *"Men of Authority": The New Zealand Maori Council and the Struggle for Rangatiratanga in the 1960s-1970s*. Treaty of Waitangi Research Unit. Rangatiratanga Series. Victoria University: Wellington, 2007.

Butterworth, G and Butterworth, S. *Reforming Education. The New Zealand Experience 1984–1996*, Palmerston North: Dunmore Press, 1998.

Campbell, A.E. *The Control of Post-primary Schools: A Report on an Enquiry made in the Auckland District*, Wellington: NZCER, 1948.

Capper, P. "Wrong Questions—Wrong Answers." *PPTA Journal*, 3, 1988, 10–13.

Centre for Contemporary Studies, *Unpopular Education. Schooling and Social Democracy in England since 1945*. London: Hutchinson, 1981.

Chilton, P. "Missing Links in Mainstream CDA," in *A New Agenda in (Critical) Discourse Analysis*, edited by R. Wodak and P. Chilton, 19–51. Amsterdam & Philadelphia: J. Benjamins, 2005.

Clark, J. "The Strange Case of the Ministry of Education's Mysterious Philosophy of the Curriculum," *Delta*, 51, no. 1, 2000, 41–53.

Cocklin, B. "Separate or Mixed Schooling: A Revisionist Study of Secondary Education in Marlborough 1946–58." MA diss. Palmerston North: Massey University, 1983.

Codd, J. "Picot: a Risky Reform?" *PPTA Journal*, 3, 1988, 2–5.

———. "Democratic Principles and the Politics of Curriculum Change in New Zealand." In *The Politics of Education in New Zealand*, edited by M. Clark, 43–65. Wellington: NZCER, 1981.

———. "Educational Reform, Accountability and the Culture of Distrust." *New Zealand Journal of Educational Studies*, 34, no. 1, 1999, 45–53.

Codd, J. and Gordon, L. "School Charters: The Contractualist State and Education Policy." *New Zealand Journal of Education Studies*, 26, no. 1, 1991, 21–34.

Codd, J., Harker, D., and Nash, J. (eds.). *Political Issues in New Zealand Education*. Palmerston North: Dunmore Press, 1985.

Cohen, S. and Young, J. (eds.). *The Manufacture of News. Deviance, Social Problems and the Mass Media*. London: Constable, 1981.

Coons, J.E. and Sugarman, S.D. *Education by Choice: The Case for Family Control.* University of California Press, 1978.

Crocombe, G., Enwright, M.J., and Porter, M.E. *Upgrading New Zealand's Competitive Advantage.* Oxford University Press: Auckland, Melbourne, Oxford and New York, 1991.

Dalziell, A. "Tomorrow's Schools from a Rural Perspective," *PPTA Journal*, 3, 1988, 21–23.

Dawkins, R. *The God Delusion.* London: Bantam Press, 2006.

Down, B.F.R. "Re-reading the History of Western Australian State Secondary Schooling after 1945." PhD diss. Murdoch University, 1993.

Dunstall, G. "The Social Pattern." In *The Oxford History of New Zealand*, edited by G.W. Rice, 451–480. Second edition. Auckland: Oxford University Press.

Easton, B. "The Restructuring and Liberalisation of the New Zealand Economy." *British Review of New Zealand Studies*, 6, November 1993, 75–92.

Education Policy Response Group Occasional Paper. *Educational Reform in New Zealand 1989–1999. Is There Any Evidence for Success?* Palmerston North: Massey University College of Education, 1999.

Edwards, B. and Moore, J. "Hegemony and the Culturalist State Ideology in New Zealand," in *The Politics of Confirmity*, edited by R. Openshaw and E. Rata, Auckland: Pearson Education, 2009 forthcoming.

Ellison, P. *The Manipulation of Maori Voice: A Kaupapa Maori Analysis of the Picot Policy Process.* Department of Education. Institute for Research and Development in Maori Education. Victoria University, Wellington, 1997.

Endres, A.M. "The Political Economy of W.B. Sutch: Toward a Critical Appreciation," *New Zealand Economic Papers*, 20, no. 1, 1986, 17–40.

Ennis, J. "How Failure Begins in the Classroom." *Tu Tangata*, 36, June/July 1987, 22–23.

———. "Why Do Our Schools Fail the Majority of Maori Children?" *Tu Tangata*, 36, June/July 1987, 21–22.

Evans, R. *Defence of History.* London: Granta Books, 1997.

Fairclough, N. "Critical Discourse Analysis in Transdisciplinary Research." In *A New Agenda in (Critical) Discourse Analysis*, edited by R. Wodak and P. Chilton, 53–70. Amsterdam & Philadelphia, 2005.

"First Things First," *PPTA Journal*, 9, no. 8, 1962, Editorial.

Fitzsimons, P., Peters, M., and Roberts, P. "Economics and the Educational Policy Process in New Zealand." *New Zealand Journal of Educational Studies*, 34, no. 1, 1999, 35–44.

Freeman-Moir, J. "Enjoyable and Quiet: The Political Economy of Human Misdevelopment." *New Zealand Journal of Educational Studies*, 16, no. 1, May 1981, 15–27.

Franklin, S.H. *Trade Growth and Anxiety. New Zealand beyond the Welfare State.* Wellington: Methuen, 1978.

Fry, R. "Feature: School and Community Partnership." *PPTA Journal* 2 (1988), 1.

———. Editorial. *PPTA Journal*, 3, 1988, 1.

———. *Its Different for Daughters. A History of the Curriculum for Girls in New Zealand Schools, 1900–1975.* Wellington: NZCER, 1985.

Gibbons, P.J. "The Climate of Opinion." In *The Oxford History of New Zealand*, edited by G.W. Rive, 308–336. Auckland: Oxford University Press, 1992.

Gilbert, J. *Catching the Knowledge Wave? The Knowledge Society and the Future of Education.* Wellington: NZCER Press, 2005.

Goldsmith, P. (2004). "Review of *Treasury: The New Zealand Treasury, 1840–2000* by Malcolm McKinnon, Auckland University Press: Auckland, 2003." *New Zealand Journal of History*, 38, no. 1, 2004, 82–83.

Gordon, E.A. "Access. The Limits and Capacity of the State." DPhil. diss., Palmerston North: Massey University, 1989.

———. "Picot and the Disempowerment of Teachers." *Delta*, 41, May 1989, 23–32.

Gordon, L. and Codd, J. *Education Policy and the Changing Role of the State.* Proceedings of the New Zealand Association for Research in Education Seminar on Education Policy, Massey University, July 1990, *Delta Studies in Education*, Number 1, Massey University: Palmerston North, 1990.

Gouldner, A. *The Future of Intellectuals and the Rise of the New Class.* London: MacMillan, 1979.

Grace, G. "Labour and Education: The Crisis and Settlements of Education Policy." In *The Fourth Labour Government. Radical Policies in New Zealand.* Second edition, edited by M. Holland and J. Boston, 165–191. Auckland: Oxford University Press, 1990.

———. "The New Zealand Treasury and the Commodification of Education. In *New Zealand Education Policy Today: Critical Perspectives*, edited by S. Middleton, J. Codd and A. Jones, 27–39. Wellington: Allen and Unwin, 1990.

Hanna, A. "Teachers Answer Back." *New Zealand Commerce*, 3, no. 4, October 15, 1947, 25.

Hawke, G. "Economic Trends and Economic Policy, 1938–1992." In *The Oxford History of New Zealand*. Second edition, edited by R.W. Rice, 412–450. Auckland: Oxford University Press, 1992.

Hingangaroa-Smith, G. "The Picot Report: A Cocktail for a Cultural and Social Catastrophe." *PPTA Journal*, 3, 1988, 17–18.

Hobsbawm, E. and Ranger, T., ed. *The Invention of Tradition.* Cambridge: Cambridge University Press, 1983.

Hummel, R.P. *The Bureaucratic Experience.* Third edition. New York: St Martins Press, 1987.

Harker, D. "Schooling and Cultural Reproduction." In *Political Issues in New Zealand Education.* Edited by J. Codd, R. Harker, and R. Nash, 57–72. Palmerston North: Dunmore Press, 1985.

Hunter, J.D. and Fessenden, T. "The New Class as a Capitalist Class: The Rise of the Moral Entrepreneur in America," In *Hidden Technocrats, the New Class and New Capitalism*, edited by H. Kellner, H. and F.W. Heuberger. London: Transaction Publishers, 1992.

James, C. *New Territory. The Transformation of New Zealand, 1984–92*, Wellington: Bridget Williams, 1992.

———. The *Quiet Revolution. Turbulence and Transition in Contemporary New Zealand*, Wellington: Allen and Unwin, 1986.

Jencks, C. *A Report on Financing Education by Payments to Parents*, Cambridge, MA: Center for the Study of Public Policy, 1970.

Jenkins, K. *Rethinking History*. London and New York: Routledge, 1991.

Jesson, J.G. "The PPTA and the State: From Militant Professionals to Bargaining Agent. A Study of Rational Opportunism." PhD diss. University of Auckland, 1995.

Joshua, H. and Wallace, T. *To Ride the Storm. The 1980 Bristol "Riot" and the State*. London: Heineman Educational Press, 1983.

Kenworthy, L.M., Martindale, T.B., and Sadaraka, S,M. *Some Aspects of the Hunn Report. A Measure of Progress*. Wellington: Victoria University, School of Political Science and Public Administration, 1968.

Kerr, R. "Stand and Deliver. A Business Leader Challenges the Education Establishment to Enter the Real World." *Metro*, 8, no. 129, March 1992, 110–115.

Keesing, R.M. "Creating the Past: Custom and Identity in the Contemporary Pacific." *The Contemporary Pacific*, nos. 1 and 2, 1989, 19–42.

Kuper, A. *Culture. The Anthropologists' Account*. Cambridge, MA, and London: Harvard University Press, 2001.

Lankshear, C. "The Picot Report as a Cultural Nightmare." *PPTA Journal*, 3, 1988, 6–9.

Lauder, H. *The Lauder Report. Tomorrow's Education, Tomorrow's Economy*. A report commissioned by the Education Sector Standing Committee of the New Zealand Council of Trade Unions: Wellington, June 1991.

Lauder, H., Middleton, S., Boston, J., and Wylie, C. "The Third Wave; a Critique of the New Zealand Treasury's Report on Education." *New Zealand Journal of Educational Studies*, 23, no. 1, 1988, 15–43.

Lee, G.D. "From Rhetoric to Reality: A History of the Development of the Common Core Curriculum in New Zealand Post-primary Schools." PhD diss. University of Otago, 1991.

Lee, G. and Hill, D. "Curriculum Reform in New Zealand: Outlining the New or Restating the Familiar?" *Delta*, 48, no. 1, 1996, 19–32.

Lee, G., Hill, D., and Lee, H. "The New Zealand Curriculum Framework: Something Old, Something New," In *Reshaping Culture, Knowledge and Learning? Policy and Content in the New Zealand Curriculum Framework*, edited by A.-M. O'Neill, J. Clark, and R. Openshaw, 71–89. Palmerston North: Dunmore Press, 2004.

Lee, G., Lee, H., and Openshaw, R. "The Comprehensive Ideal in New Zealand: Challenges and Prospects." In *The Death of the Comprehensive High School? Historical, Contemporary, and Comparative Perspectives*, edited by B.M. Franklin and G. McCulloch, 169–184. New York and Basingstoke, UK: Palgrave Macmillan, 2007.

Lee, H.F. "The Credentialed Society: A History of New Zealand Public School Examinations 1871–1990." PhD diss. University of Otago, 1991.

Lee, H., O'Neill, A.-M., and McKenzie, J. "'To Market, To Market....' The Mirage of Certainty: An Outcomes-based Curriculum." In *Reshaping Culture, Knowledge and Learning? Policy and Content in the New Zealand Curriculum*

Framework, edited by A.-M. O'Neill, J. Clark, and R. Openshaw, 47–70. Palmerston North: Dunmore Press, 2004.
Levett, A. "Education Reviews Fall Short." *National Business Review*, July 3, 1988, 6.
Maori Synod of The Presbyterian Church, *A Maori View of the Hunn Report*, Christchurch and Whakatane: Presbyterian Bookroom and Te Waka Karaitiana Press, 1961.
Marshall, G.N. "The development of secondary education in New Zealand from 1935 to 1970." PhD diss. Hamilton: University of Waikato, 1989.
Matthews, M. *Challenging New Zealand Science Education*. Palmerston North: Dunmore Press, 1995.
Metge, J. *The Maoris of New Zealand*. London: Routledge and Kegan Paul, 1967.
———. *The Maoris of New Zealand*. Revised edition. London: Routledge & Kegan Paul, 1976.
McCulloch, G. "Historical Perspectives on New Schooling." In *Myths and Realities. Schooling in New Zealand*, edited by A. Jones, G. McCulloch, J. Marshall, H.S. Smith, and L.T. Smith, Palmerston North: Dunmore Press, 1990.
———. "Secondary Education Without Selection? School Zoning Policy in Auckland since 1945." *New Zealand Journal of Educational Studies*, 21, no. 2, November 1986, 98–112.
———. "Serpent in the Garden: Conservative Protest, the 'New Right' and New Zealand Educational History." *History of Education Review*, 20, no. 1, 1991, 73–97.
———. "'Spens v. Norwood': contesting the Educational State?" *History of Education*, 22, no. 2, 1993, 163–180.
McKenzie, D., Lee, H., and Lee, G. *Scholars or Dollars? Selected Historical Case Studies of Opportunity Costs in New Zealand Education*, Palmerston North: Dunmore Press, 1996.
McKinnon, M. *Treasury: The New Zealand Treasury, 1840–2000*. Auckland University Press: Auckland, 2003.
McLaren, I.A. *Education in a Small Democracy: New Zealand*. World Education Series. London: Routledge and Kegan Paul, 1974.
Meikle, P. *School and Nation. Post Primary Education since the War*. Wellington: New Zealand Council for Educational Research, 1961.
Middleton, M. "American Influences on New Zealand Sociology of Education, 1950–1988." In *The Fulbright Seminars. The Impact of American Ideas on New Zealand's Educational Policy, Practice and Thinking*, edited by J. Philips, J. Lealand, and G. MacDonald, 50–55. New Zealand–United States Educational Foundation, Wellington: NZCER, 1989.
Middleton, S. "Sexual Apartheid or Androgyny? Four Contemporary Perspectives on Women's Education in New Zealand." *New Zealand Journal of Educational Studies*, 17, no. 1, May 1982, 57–73.
———. "Family Strategies of Cultural Reproduction." In *Political Issues in New Zealand Education*, edited by J. Codd, R. Harker, and R. Nash, Palmerston North: Dunmore Press, 1985, 83–100.

Middleton, S. "Feminism and Education in Post-War New Zealand." In *Reinterpreting the Educational Past*, edited by R. Openshaw and D. McKenzie, 132–146. Wellington: NZCER, 1987.

Miller, P. *Long Division. State Schooling in South Australian Society.* Netley, South Australia: Wakefield Press, 1986.

Mintrom, M. and Wanna, J. "Innovative State Strategies in the Antipodes: Enhancing the Ability of Governments to Govern in the Global Context." *Australian Journal of Political Science* 41, no. 2, June 2006, 161–176.

Moe, T.M. *Schools, Vouchers, and the American Public.* Washington, DC: Brookings Institution Press, 2001.

Moss, L. "Picot as Myth." In *Critical Perspectives: New Zealand Education Policy Today*, edited by S. Middleton, J. Codd, and A. Jones, 139–149. Wellington: Allen and Unwin in association with Port Nicholson Press, 1990.

Muldoon, R.D. "The Future of University Education in New Zealand," *Delta*, 2, May 1968, 5. This article also includes an appended critical commentary by *Delta* editor, R.J. Bates.

Mundy, P.J. "The Current Crisis in New Zealand Education," *Delta*, 2, May 1968, 29–33.

Murdoch, J.H. *The High Schools of New Zealand. A Critical Survey.* Christchurch: NZCER, 1944.

Nash, R. "The Schooling Debate in New Zealand: Some British Parallels." *New Zealand Journal of Educational Studies*, 15, no. 2, November 1980, 117–126.

———. "Society and Culture in New Zealand: An Outburst for 1990." *New Zealand Sociology*, 5, no. 2, 1990, 99–124.

———. "The Treasury on Education: Taking a Long Spoon." *New Zealand Journal of Educational Studies*, 23, no. 1, 1988, 35–43.

Nash, R. and Korndorffer, W. "After Picot Community Education Will Be Something Else." *PPTA Journal*, 2, 1988, 5–9.

New Zealand Council for Educational Research, *How Fair Is New Zealand Education?* Part 1. NZCER for Royal Commission on Social Policy, Wellington, 1987.

New Zealand Department of Education. *The Curriculum Review.* Report of the Committee to Review the Curriculum for Schools, Wellington, 1987.

———. *Education Policies in New Zealand: Report Prepared for the OECD Examiners by the Department of Education in March 1982*, Wellington, 1983.

———. OECD Review of New Zealand Educational Policies 1982. Draft Statement for the Examining Panel. Department of Education. Wellington, February 1982.

———. *The Secondary School Curriculum 3. The Challenge Is Change.* Wellington: Government Printer, 1972.

New Zealand Ministry of Education. *Education for the Twenty-First Century.* Wellington: Learning Media, 1994.

———. *The New Zealand Curriculum Framework.* Wellington: Learning Media, 1993.

———. White Paper entitled, *The National Qualifications Framework of the Future.* Wellington: Ministry of Education, 1999.

New Zealand Post-Primary Teachers' Association. *Education in Change. Report of the Curriculum Review Group*. Auckland: Longman Paul, 1969.

New Zealand Qualifications Authority. *An Introduction to NCEA*. Wellington: NCEA-001 B, c. 2002.

———. *Learning to Learn. An Introduction to the New Zealand Qualifications Framework*, 1992.

———. *NCEA. Better Information for Employers*, Wellington: New Zealand Qualifications Authority. Fold-over pamphlet, c. 2002.

O'Brien, J. *A Divided Unity! Politics of NSW Teacher Militancy since 1945*. Sydney: Allen and Unwin in association with New South Wales Teachers' Federation, 1987.

Olssen, M. "Restructuring New Zealand Education: Insights from the Work of Ruth Jonathan." *New Zealand Journal of Educational Studies*, 34, no. 1, 1999, 54–65.

O'Neill, A.-M. "Reading Educational Policy." MEd diss., Palmerston North: Massey University, 1993.

Openshaw, R. *Between Two Worlds. A History of Palmerston North College of Education, 1956–1996*. Palmerston North: Dunmore Press, 1996.

———. "Citizen Who? The Debate over Economic and Political Correctness in the Social Studies Curriculum." In *New Horizons for New Zealand Social Studies*, edited by P. Benson and R. Openshaw, 19–42, Massey University: ERDC Press, 1998.

———. "Diverting the Flak: the Response of the New Zealand Department of Education to Curriculum Controversy." *Change. Transformations in Education* 4, no. 1, May 2001, 33–47.

———. "'Nothing Objectionable or Controversial': The Image of Maori Ethnicity and 'Difference' in New Zealand Social Studies." In *Struggles over Difference. Curriculum, Texts, and Pedagogy in the Asia-Pacific*, edited by Nozaki, Y., Openshaw, R. and Luke, 25–40. A. Albany, New York: State University of New York Press, 2005.

———. "Preparing for Picot: Revisiting the 'Neoliberal' Education Reforms". *New Zealand Journal of Educational Studies*, 38, no. 2, 2003, 135–150.

———. "Putting Ethnicity into Policy. A New Zealand Case Study." In *Public Policy and Ethnicity. The Politics of Ethnic Boundary-Making*, edited by E. Rata and R. Openshaw, 113–127, Hampshire and New York: Palgrave Macmillan, 2006.

———. "Review of Jane Gilbert: *Catching the Knowledge Wave? The Knowledge Society and the Future of Education*." *New Zealand Journal of Psychology*, 2007.

———. "Schooling in the 40's and 50's: An Oral History." *Research Resources*, no. 1. Massey University, Palmerston North: ERDC Press, 1991.

———. "Subject Construction and Reconstruction; Social Studies and the New Mathematics." In *The School Curriculum in New Zealand: History, Policy and Practice*, edited by G. McCulloch, 201–208, Palmerston North: Dunmore Press, 1992.

———. "Upholding BasicValues: A Case Study of a Conservative Pressure Group." In *Political Issues in New Zealand Education, edited by* J. Codd, R. Harker and R. Nash, 231–147. Palmerston North: Dunmore Press, 1985.

Openshaw, R. *Unresolved Struggle. Consensus and Conflict in State Post-primary Education*. Palmerston North: Dunmore Press, 1995.

Openshaw, R., Lee H., and Lee, G. *Challenging the Myths. Rethinking New Zealand's Educational History*. Palmerston North: Dunmore Press, 1993.

Openshaw, R. and Walshaw, M. "Pandering to Women? The Introduction of the Science and Mathematics New Scheme in an Era of Ambiguity." *New Zealand Journal of Educational Studies*, 41, no. 2, 2006, 241–255.

Openshaw, R, Lee, G., and Lee, H. "The Comprehensive Ideal in New Zealand Education: Challenges and Prospects." In *Death of the Comprehensive High School: Historical, Contemporary and Comparatives Perspectives*, edited by B.M. Franklin and G. McCulloch, 169–184. New York and London: Palgrave Macmillan, 2008.

Parsonage, W. "The Education of Maoris in New Zealand." *Polynesian Society Journal* 65, no. 1, 1956, 5–11.

Peachey, A. *What's Up With Our Schools? A New Zealand Principal Speaks Out*. Auckland: Glenfield: Random House, 2005.

Pearson, D.C. and Thorns, D.C. *Eclipse of Equality: Social Stratification in New Zealand*. Sydney: Allen and Unwin, 1983.

Penetito, W. "Taha Maori and the Core Curriculum." *Delta*, 34, July, 1984, 34–43.

Picot, B. "New Zealand Attitudes to Challenge and Change. A Discussion Paper." Auckland: Brian Picot, March 1979.

———. *South Africa's Apartheid. An Attempt at Perspective*. Auckland: Brian Picot, November 1980.

———. *Working Together*. Wellington: New Zealand Planning Council. NZPC no. 7, June 1978.

Piddington, R. An Introduction to Social Anthropology, 1 and 2. Oliver & Boyd: Edinburgh, 1950, 1957, 776.

Polanyi, K. *The Great Transformation*. New York: Farrar and Rinehart, 1944.

Powell, G., "The Maori School. A Cultural Dynamic." *Polynesian Society Journal*, 64, no. 3, 1955, 259–266.

PPTA Executive. "Tomorrow's Schools: Yesterday's Mistake?" A paper presented to the PPTA Annual Conference. Wellington, September 30–October 2, 2008.

PPTA Journal, 1, no. 8, November 1955, editorial.

"Primary Education an Easy Target." *National Education*, 70, no. 3, August 1988, 114–117. No author given.

Ramsay, P. "The Domestication of Teachers. A Case of Social Control." In *Political Issues in New Zealand Education, edited by* J. Codd, R. Harker and R. Nash, 103–122. Palmerston North: Dunmore Press, 1985.

Ramsay, P., Sneddon, D.G., Grenfell, J., and Ford, I. *Tomorrow Will Be too Late: Schools with Special Needs in Mangere and Otara: Final Report of the Schools with Special Needs Project*. Hamilton: Education Department, University of Waikato, 1981.

Rata, E. "'Goodness and Power': The Sociology of Liberal Guilt." *New Zealand Sociology*, 11, no. 2, November 1996, 223–276.

Rata, E. and Openshaw, R., eds. *Public Policy and Ethnicity. The Politics of Ethnic Boundary-Making*. London: Palgrave Macmillan, 2006.
Ravitch, D. *The Revisionists Revised. A Critique of Radical Revisionism*, New York: Basic Books, 1977.
Ritchie, J. *Child Rearing Patterns in New Zealand*. Auckland: A.H. and A.W. Reed, 1970.
Robbins, T. *Cults, Converts and Charisma: the Sociology of New Religious Movements*. London: Sage, 1988.
Roulston, D. "Educational Policy Change, Newspapers and Public Opinion in New Zealand, 1988–1999." *New Zealand Journal of Educational Studies*, 41, no. 2, 2006, 145–162.
———. "An Historical Study of Newspapers and Public Opinion about Education." MA diss., Wellington: Victoria University, 1986.
Ryan, A. "'For God, Country and Family.' Popular Moralism and the New Zealand Moral Right." MA diss., Palmerston North: Massey University, 1986.
Salmond, A. *Hui. A Study of Maori Ceremonial Gatherings*. Wellington: A.H. and A.W. Reed, 1975.
Salter, B. and Tapper, T. *Education, Politics and the State*. London: Grant McIntyre, 1981.
Sandall, R. *The Culture Cult. Designer Tribalism and Other Essays*. Boulder, Colorado: Westview Press, 2001.
Savage, G. "Social Class and Social Policy: The Civil Service and Secondary Education in England During the Interwar Period." *Journal of Contemporary History*, 7, no. 18, 1983, 261–280.
Schwimmer, E.G., ed. *The Maori People in the Nineteen Sixties*. Auckland: Blackwood and Janet Paul, 1968.
Scott, D.J. "The Currie Commission and Report on Education in New Zealand 1960–62." PhD diss. University of Auckland, 1996.
Scott, N. *Survey of Some Aspects of New Zealand Secondary Schools*. Wellington. Department of Education, 1977.
Shuker, R. The *One Best System? A Revisionist History of State Schooling in New Zealand*. Palmerston North: Dunmore Press, 1987.
Shuker, R. and Openshaw, R. *Youth, Media and Moral Panic in New Zealand*. Delta Research Monograph no. 11. Palmerston North: Massey University, 1990.
Simon, B. "The 1944 Education Act: A Conservative Measure?" *History of Education* 15, no. 1 (1986), 31–43.
Smith, G.H. and Smith, L.T. "Ki Te Whai Ao, Ki Te Ao Marama: Crisis and Change in Maori Education," in *Myths and Realities. Schooling in New Zealand*, edited by A. Jones, G. McCulloch, J. Marshall, H.S. Smith and L.T. Smith, 123–156, Palmerston North: Dunmore Press, 1990.
Snook, I. "Educational Reform in New Zealand: What *Is* Going On?" *Access*, 8, no. 2, 1989, 9–18.
———. "Policy Change in Higher Education. The New Zealand Experience," *Higher Education*, 21, 1991, 621–634.

Soler, J. "Reading the Word and the World: the Politics of the New Zealand Primary School Literary Curriculum from the 1920s to the 1950s." PhD diss. University of Otago, 1996.

Spoonley, P. *The Politics of Nostalgia. Racism and the Extreme Right in New Zealand.* Palmerston North: Dunmore Press, 1987.

Spring, J. *Education and the Rise of the Corporate State*, Boston: Beacon Press, 1972. Foreword by Ivan Illich.

Sutch, W.B. *The Quest for Security in New Zealand.* London: Oxford University Press, 1966.

———. *The Maori Contribution—Yesterday, Today and Tomorrow.* Wellington: Department of Industries and Commerce, 1964.

Sweetman, R. *A Fair and Just Solution? A History of the Integration of Private Schools in New Zealand*, Palmerston North: Dunmore Press in association with the Ministry for Culture and Heritage and the Association of Proprietors of Integrated Schools, 2002.

"The Parry Report." Editorial, *PPTA Journal*, 6, no. 2, March 1959, 1.

The World Atlas 1994. Washington: World Bank.

Thomson, D. *Selfish Generations? The Ageing of New Zealand's Welfare State*, Wellington: Bridget Williams, 1991.

Thrupp, M. Introduction to *New Zealand Journal of Educational Studies*, 34, no. 1, 1999, 5–7.

Velde, M. "Better Outcomes", *New Zealand Education Gazette. Tukutuku Korero* 79, no. 10, June 5, 2000, 1.

Vellekoop, C. "Migration Plans and Vocational Choices of Youth in a Small New Zealand Town," *New Zealand Journal of Educational Studies*, 3, no. 1, 1969, 10–40.

———. "Streaming and Social Class," Delta, 5, August 1969, 12–21.

Walker, R. "Cultural Domination of Taha Maori. The Potential for Radical Transformation," in *Political Issues in New Zealand Education*, edited by J. Codd, R. Harker and R. Nash, 73–82, Palmerston North: Dunmore Press, 1985.

———. "The Modern Maori Warden." *Te Maori*, 2, July–August 1971, 33–37.

———. "Maori People since 1960." In *The Oxford History of New Zealand.* Second edition, edited by G.W. Rice, 498–519. Auckland: Oxford University Press, 1992.

Wallace, J.N. "Presidential Address to 1961 Annual Conference." *PPTA Journal*, 8, no. 9, October 1961, 14–17.

Watson, H. "The Fate of Women and Girls in Tomorrow's Schools," *PPTA Journal* 3 (1988), 29–30.

Webb, L. *The Control of Education in New Zealand*, Auckland: NZCER, 1937.

Webster, B.A. "The Politics of the New Zealand Post-Primary Teachers' Association." In *The Politics of Education in New Zealand*, edited by M. Clark, 191–206. Wellington: NZCER, 1981.

Webster, S. *Patrons of Maori Culture. Power, Theory and Ideology in the Maori Renaissance.* Dunedin: University of Otago Press, 1998.

Wellington, M. *New Zealand Education in Crisis*. Auckland: Endeavour Press, 1985.
Whitehead, C. "The Thomas Report: A Study in Educational Reform." *New Zealand Journal of Educational Studies*, 9, no. 1, 1974, 54–64.
Wood, C.J. "Education as a Preparation for Employment." *New Zealand Commerce*, 3, no. 4, 15 October, 1947, 24–27.
Wrong, D.H. "Cultural Relativism as Ideology." *Critical Review*, 11, 1997, 291–300; 292–293.
Wylie, C. *The Impact of Tomorrows' Schools in Primary Schools and Intermediates. 1991 Survey Report*. Wellington: NZCER, 1992.
Yee, B. "Women Teachers in the Primary Scene: A Study of their Access to Power and Decision-making." MA diss. University of Canterbury, 1985.

Index

A Nation at Risk, 84–85
Access, 107–108
Administering for Excellence (The Picot Report)
    place in history, 3–18, 165–184
    press reaction to, 126–127
    release of, 3, 4, 118
    submissions to, 130–131
    writing of, 99–122
    see also Tomorrows', School, and Taskforce (Picot)
assessment and examinations
    Proficiency, 20
    School Certificate and University Entrance, 20–21, 25–28, 33, 49
    NCEA, 27, 162–164, 172
    NZQA, 49, 156–157

Bazley, Margaret, 117, 179
Beeby, C. E., 21, 24, 53, 109
Benton, Richard, 109, 114–115
Boards of trustees and school committees, 4, 28, 47, 131–136, 167–168
bulk funding
    in *Tomorrow's schools*, 139
    attempted introduction under National, 157–161

Cartwright Report, 91, 108, 177, 179
Citizen's Association for Racial Equality (CARE), 70, 73, 109
Committee of Inquiry into Curriculum, Assessment and Qualifications, 92
    see also Marshall, Russell
Committee of Officials on Public Expenditure (COPE), 86–89
community
    concept of, 47, 114–115, 178
    Community Education Forums, 4, 159
Community Education Initiative Scheme (CEIS), 71–72, 178
Critical Discourse Analysis (CDA), 180–181
critical theory, 9–15, 170–181
Crowther Report (UK), 35–36
core curriculum, 21–2, 24
Curriculum Review (1987), 70, 81–83, 106, 156
Currie Report, 33–39, 178

devolution, 9, 74, 104–106, 115, 118, 131
Douglas, Roger, 99, 137

economic factors (general)
    crises and responses, 13, 39, 43–58, 146–147
    historic trends, 20–21, 39, 86
education boards
    abolition of, 4, 120
    criticism of, 112

education boards—*Continued*
  function of, 30
  regrets at abolition of, 120
educational consensus, 8, 19
Education Development Conference (EDC), 46, 76–77, 102, 174
Education Policy Council, 119–120
Education Review Office (ERO), 4, 158

Fargher-Probine Report, 94, 102–104, 177
feminism/feminists
  differentiation, 20–22
  influence of, 15, 54–56, 179
Franklin, S. Harvey, 31, 45–46, 178

Gianotti, Maurice, 100–102
Gibbs Report, 91, 127
global context of reforms, 5, 43–44, 173–183

Holmes, Sir Frank, 4, 31, 38, 43, 86, 178–179

ideologies influencing reform
  academic radicalism, 51–54
  moral right, 56–58
  see also feminism/feminist, and neoliberalism/new right

Labour Governments prior to 1984
  First Labour Government, 8, 19–20
  Second Labour Government, 50–51, 86
Labour Government (Clark), 161–163
Lange, David (Labour prime minister)
  concerns over reforms, 3, 13, 104, 130
  defense of reforms, 123–124, 130
  expectations of reforms, 82–85, 110–111
  Lange papers, 17, 125
  place in history, 130, 137, 176

Logos (public relations firm), 17, 125–128, 131, 176
Lough Report, 145–146

Māori
  activism and agency, 15, 59–74, 113, 116, 120, 174
  biculturalism, 59–74, 120, 174
  disadvantage, 22–23, 107–109, 167
  Taha Māori, 68–69
  reaction to reforms, 132–133, 139
  see also Te Kohanga Reo
  see also National Advisory Committee on Māori Education
Marshall, Russell, 80–82, 99–110
McCombs Report, 77–78, 174
National Advisory Committee on Māori Education (NACME), 61, 109

National Government (Bolger), 147–161
National Governments prior to 1990, 23, 29–30, 50, 86, 147–161
neoliberalism/new right, 8–11, 107–110, 137, 147, 176–180
new policy discourse, 90–95, 181–183
New Zealand Curriculum Framework (NZCF), 16, 154–156, 170–171
New Zealand Department of Education
  abolition of, 4, 110
  criticism of, 30, 77, 93–4, 111–112
  defense of secondary education, 2–5, 22–31, 33–34, 50–51
  functions, 3, 101
  links with PPTA, 50
  reactions to Māori activism, 61–74
  response to pressure groups, 53–58
  submission to Picot taskforce, 110
  under pressure from Treasury, 85–89, 175–176

Index

see also New Zealand Ministry of Education and Renwick, W. L.
New Zealand Ministry of Education
  continuity with former Department, 144
  creation of, 4, 11, 111, 143–145
  pressure on, 145–152
New Zealand Post-Primary Teachers' Association (NZPPTA), 47–51, 83, 122, 128–129, 131, 140–143, 156, 164, 168
New Zealand Secondary School Boards Association, 83, 131
New Zealand Planning Council, 45–46, 87, 99
Nordmeyer Report (1974), 76, 102–103

Organisation for Economic Cooperation and Development (OECD), 10, 78–79, 83–85, 123, 174

Parent Advocacy Council, 4, 120
parental choice, 4, 25, 54, 121
Picot, Brian
  background, 99
  publications, 46–47
  taskforce leadership, 4, 99–122
  views of, 46–47, 103–104, 175
  see also Administering for Excellence, and Taskforce (Picot)
Polanyi, Karl, 114–115, 178
Porter Project, 6, 29, 36, 146, 152–154
Prime Minister's Department, 17, 126, 128, 178
propaganda war, 123–142, 176–177

Ramsay, Peter (Taskforce member), 54, 100, 119, 137
reinterpreting the reforms, 3–18, 165–184
Renwick, W.L. (Director-General of Education), 51, 55, 63, 73, 87–89, 181

Richardson, Ruth, 80–81, 92, 150
  see also Nation at Risk
Rosemergy, Margaret (Taskforce member), 100
Royal Commission on Social Policy, 83, 94–95, 127

school charters, 70, 77, 82, 119, 144–145, 160
school–labour market relationships, 19, 23–25, 56, 83
  see also economic factors (general)
Scott Report (1986), 83, 92–94, 108
secondary education
  academic standards, 19–39
  criticism of, 28–39, 43–58, 59–74, 75–95, 108
Smith, Lockwood, 130, 147, 149, 151, 154–155, 158, 160–161
Sexton Report, 6, 29, 148
Smelt, Simon, 5, 6, 106–107, 109
State Services Commission (SSC)
  influence on reforms, 45, 106, 115–117, 174–175
  pressure on Ministry of Education, 145–152, 174–175
  sources for, 18

Te Kohanga Reo, 23, 69–70, 106–108, 112–113, 133, 178
Thatcher, Margaret (British prime minister), 10, 43, 109, 178
Thomas Report (1943), 21–24, 28, 36, 49
Taskforce (Picot)
  brief, 85, 95, 100–101
  disagreements among members, 117
  influences on, 18, 74, 99–122
  initial meetings, 101–106
  recommendations, 4, 43
  report of, 3
  setting up of, 3, 14–15, 19, 83

Tomorrow's Schools (1988)
  general, 4, 8, 11, 15–16, 23,
    68, 74, 129,
    143, 178
  cementing of ideals, 161–164,
    183–184
  evidence for success of, 165–173
  provisions of, 138–139
  reaction to, 140–142
  release of, 129, 137
Treasury
  changes in, 38, 44, 85
  Government Management: Brief to the Incoming Government (1987), 8, 107–110
  influence on reforms, 6, 9–11, 14, 45, 115–117, 158–159, 175
  pressure on Department of Education, 85–89
  pressure on Ministry of Education, 145–152, 174–175
  sources for, 18

Victorian educational reforms, 116, 118
vouchers, 112–113, 178

Waitangi Tribunal, 73
Webb, Leicester, 105, 107, 184
Wereta, Whetu (Taskforce member), 100, 104, 119, 139
Wise, Colin (Taskforce member) 100, 104

zoning, 24–25, 121, 159

GPSR Compliance
The European Union's (EU) General Product Safety Regulation (GPSR) is a set of rules that requires consumer products to be safe and our obligations to ensure this.

If you have any concerns about our products, you can contact us on

ProductSafety@springernature.com

In case Publisher is established outside the EU, the EU authorized representative is:

Springer Nature Customer Service Center GmbH
Europaplatz 3
69115 Heidelberg, Germany

www.ingramcontent.com/pod-product-compliance
Lightning Source LLC
LaVergne TN
LVHW051914060526
838200LV00004B/132